BANKCARD
BUSINESS

Michael J. Auriemma &
Robert S. Coley
AURIEMMA CONSULTING GROUP, INC.

AMERICAN BANKERS ASSOCIATION
1120 Connecticut Avenue, N.W.
Washington, D.C. 20036

Auriemma, Michael J.
    The bankcard business / Michael J. Auriemma and Robert S. Coley
        p. cm.
    Includes index.
    ISBN 0-89982-335-1
    1. Bank credit cards—United States. 2. Debit cards—United States. 3. Smart
    cards—United States.   I. Coley, Robert S.   II. Title.
    HG1652.U5A88   1992
    332. 1'78—dc20   92-18360
                                                                           CIP

This publication is designed to provide accurate and authoritative information in regard to the subject matter covered. It is sold with the understanding that the publisher is not engaged in rendering legal, accounting, or other professional service. If legal advice or other expert assistance is required, the services of a competent professional person should be sought.

From a Declaration of Principles jointly adopted by a Committee of the American Bar Association and a Committee of Publishers and Associations.

Contents

List of Exhibits

About the Authors

Robert S. Coley

Bob Coley has over 25 years experience in all aspects of the credit card industry.

As a managing associate with Booz, Allen & Hamilton, Coley planned, conducted, and managed bankcard implementation projects for several major issuers and associations. As vice president at Chemical Bank, his responsibilities included the management of all of Chemical's consumer credit business, the cornerstone of which was the MasterCard program. He has managed the worldwide credit and collection functions of Avis, which included all bank and T&E cards, as well as the Avis proprietary credit card.

He has assisted a major issuer establish bankcard operations in Australia, Brazil, Chile, and Argentina and negotiate the purchase and integration of bankcard portfolios in Panama and Singapore. He regularly conducts operational audits and feasibility projects for a variety of clients.

Michael J. Auriemma

Michael Auriemma has worked in the banking and credit card business for 10 years. He began his career at Chemical Bank, where his experience included branch administration, retail consumer marketing, and credit card operations.

Since 1989, he has consulted on a wide variety of projects, including organizational plans and assessments, profitability studies and modeling, processing alternative evaluations, risk assessments, and strategic alliance negotiations. He also leads the company's efforts in portfolio sales and syndicated market research projects.

Auriemma earned a BBA in business at Pace University in New York City.

Acknowledgments

We would like to express our special appreciation to Dwane Krumme, general manager and executive vice president, JCB International, whose text *Banking and the Plastic Card* served as the foundation for the book you see today. Particular thanks also go to Anita Boomstein, partner in the New York law firm of Hughes, Hubbard and Reed, for her revision of chapter 11, Legal and Regulatory Issues.

We also want to recognize the efforts of Deborah Corsi, Product Manager, and Veida Dehmlow, Managing Editor, of the American Bankers Association who have guided this book from concept to print. In addition, we extend our sincere thanks to the following members of the advisory committee who generously donated their time and suggestions in reviewing this text.

Peggy Buckwalter
Stock Transfer Agent/Dividend
 Reinvestment Administrator
Bank of Lancaster County
Lancaster, Pa.

Roney Bisio
Vice President and Manager
Mesa Grande Bank Cards
Albuquerque, N.Mex.

Tom Carotenuto
Vice President
Delaware Trust Company
Newark, Del.

Doug Baker
Management Representative
First Bank Card Center
Yankton, S.Dak.

Joan Nelson
Senior Vice President, Operations
National Bank of Blacksburg
Blacksburg, Va.

Chuck Caudle
Marketing Representative
VISA
Mclean, Va.

Al Prendergast
Senior Vice President
Human Resources
MasterCard International
New York, N.Y.

Michael Auriemma
Bob Coley
May 1992

Preface

"Charge it." That phrase is heard every day, yet until 1946 no credit card plan existed in the United States. You are about to study the fastest growing product in the history of banking—the bank credit card. Consumers have enthusiastically embraced this convenient new form of credit. From its inception as a local payment device, the bankcard has evolved into an extremely profitable—if not the most profitable—consumer product offered by the financial community. After checking accounts, bankcards have become the most widely used payment device in use today.

This book shows you how this dynamic business works. You'll follow the credit process and see the increasing importance of marketing and customer service in this hotly competitive business. Bankcards play a major role in practically every bank's consumer marketing strategy—either as a card issuer, merchant acquirer, or agent bank. Those important production functions are examined in depth, along with the quality standards that must be in place for a bankcard program to be effective and efficient.

You'll also examine the many faces of fraud that eat into profitability and look at the tools and techniques for detecting, investigating, and reporting abuses. Consumer protection laws play a major role in the bankcard business, and a chapter is devoted to the regulations that pertain.

While we tend to think primarily of credit cards (where this book's emphasis lies), other types of bankcards are also used today. You'll take a brief look at the functions of debit, ATM, check guarantee, private label, and smart cards.

Finally, you will touch on some of the strategic issues facing the industry. Much has changed since the early years of the bankcard business, and more changes lie ahead. New types of cards, applications, users, issuers, challenges, and opportunities will give rise to many questions. This is truly a dynamic business that will continue to offer excitement and reward to the successful practitioner.

1

BANKCARD HISTORY

This chapter reviews the history of bank credit cards from their rise to national prominence in the early 1960s to the state of the industry today. You will see graphic displays of the dramatic growth, gain insight into the primary factors contributing to this growth, and review the problems that arose as a result.

After successfully completing this chapter, you will be able to

♦ list and discuss the various types of bankcards

♦ outline the growth of the bankcard industry from the point of view of cardholders, merchants, and banks

♦ define the terms discount rate, country club billing, descriptive billings, and spread

♦ cite the advantages and disadvantages of revolving credit when bankcard programs were first established

Types of Cards

The past quarter of a century has seen revolutionary changes in the ways consumers bank. Plastic cards of various types have become central to the delivery of many banking services. While the focus of this text is bank credit cards, several other card types should be mentioned.

♦ **Check Guarantee Card.** This card enables merchants to accept personal checks, without risk of recourse, provided the merchant follows accepted authorization and documentation procedures. Individual banks originally provided this service in their local markets. However, today the service is offered by several large, national, nonbank organizations, most notably Telecredit and Telecheck.

recourse
The right to collect from a maker, seller, assignor, or endorser of an instrument or credit card obligation, if the first party fails to meet the obligation.

♦ **Automated Teller Machine (ATM) Card.** ATM cards are issued by individual banks to permit customers to access transaction and savings accounts 24 hours a day, every day of the year, through automated teller machines. Depending on the account relationships between issuer and customer, ATM cards may be used to withdraw funds from checking or savings accounts, make deposits, transfer funds between accounts, obtain account status information, and in some cases pay bills.

♦ **Debit Card.** Issued by a bank or other financial institution, this card permits access to a customer's checking or savings account. Debit cards are different from ATM cards in that they generally carry a MasterCard or Visa logo, thereby allowing them to be accepted at merchant locations as well as at ATMs. There has been some consumer reluctance to use debit cards because they immediately transfer funds from the user's account, thus eliminating float.

float
Money balances that appear for a period of time on both the balance statements of payer and payee due to a lag in the collection process.

♦ **Smart Card.** This card contains a computer chip with memory and interactive capability, so that data can be updated each time the card is used in an ATM or point-of-sale (POS) terminal.

♦ **T&E Card.** This card is issued by travel and entertainment companies, such as American Express and Diners Club. The principal features that distinguish these cards from bankcards are the lack of a stated credit limit and the requirement to pay the balance in full each month.

POS terminal
A device at a merchant location, which is connected to bank systems via telephone lines, used to authorize, record, and forward data electronically for each sale.

These cards and other nonbanking-related cards are discussed in greater detail in chapter 12.

The Origin of Bank Credit Cards

The origins of the bank credit card have been attributed to John C. Biggins, a consumer credit specialist at the Flatbush National Bank of Brooklyn, New York. In 1946, Biggins launched a credit plan called Charge-It. The program featured a form of scrip that was accepted by local merchants for small purchases. After the sale was completed, the merchant deposited the scrip in a bank account, and the bank billed the customer for the total scrip issued. Patterson Savings and Trust Company in New Jersey offered a similar system of credit. Not long after, in 1951, the first modern credit card was issued by the Franklin National Bank in New York.

scrip
Paper money substitute redeemable at face value at participating merchant outlets for merchandise purchased.

At about the same time, in 1950, Diners Club introduced the first travel and entertainment (T&E) card. The new card sparked the interest of banks in considering a new form of consumer lending through a plastic card that functioned similar to Biggins' scrip. This interest in credit card lending was transformed into a new lending vehicle in 1960, when Bank of America introduced BankAmericard (now Visa). The people who created this new card and the system to support it no doubt had visions of broad acceptance from the beginning. Even so, it is hard to imagine how they could have anticipated the effect bankcards would have on the purchase habits of millions of consumers throughout the world in just a few years.

Visa

BankAmericard started with a relatively small base of cardholders and merchants. The rapid growth in the number of consumers carrying cards and merchants accepting them in the years that followed would have staggered the imagination of the greatest visionary. As shown in exhibit 1.1, over 20 million consumers across the United States carried BankAmericards in 1970.

EXHIBIT 1.1   Growth of U.S. Bankcards

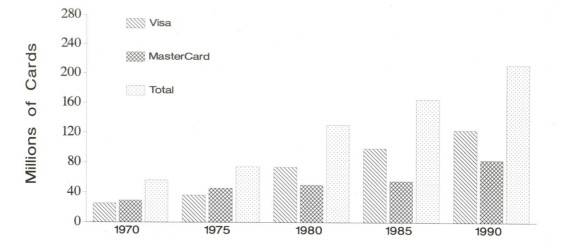

That same year, more than 170 million sales slips were processed, with each representing a single purchase. At the end of 1980, the number of cards outstanding exceeded 73 million. The number of processed sales slips reached approximately 1 billion that year. During the same 10-year period, annual gross dollar volume in the United States rose from approximately $3 billion to $31 billion, a staggering tenfold increase. At the end of 1991, Visa had 105 million accounts open, with a gross dollar volume of $171 billion. Exhibit 1.2 shows the growth over a 20-year period, a rate that is unprecedented in banking.

EXHIBIT 1.2   Bankcard Sales Volume Growth

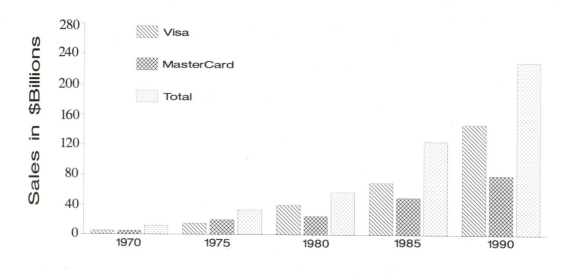

MasterCard

The rapid growth of BankAmericard to a level of national significance served as a catalyst to bring together a group of enterprising bankers, representing 17 financial institutions who were not members of BankAmericard. They decided to create a network of their own to accept one another's local credit cards. On August 16, 1966, these bankers established the Interbank Card Association (ICA) to manage the interchange functions of authorization, clearing, and settlement. An "i" symbol was placed on all cards and at merchant locations. In 1969, ICA acquired the exclusive rights to the Master Charge name and the interlocking circles design. By 1970, over 5,000 financial institutions were members of Master Charge, with nearly 36 million cardholders. As the organization continued its international expansion, the association was renamed MasterCard in 1979.

By 1980, the number of MasterCard cards in circulation in the United States had grown to over 55 million, with this number reaching 90 million by the end of 1990. Between 1980 and 1991, MasterCard U.S. gross dollar volume increased from $10.4 billion to almost $99 billion.

Reasons for Growth

Bankcards experienced such phenomenal growth over a short period for several important reasons. This section will explore those reasons from the perspectives of the cardholder, the merchant, and the bank.

Cardholders

In the early days, the issuers of credit cards faced a "chicken and egg" problem. Bankers did not know whether they should try to acquire merchants first (so that patrons would have many places to use the card) or cardholders first (so that merchants would want to accept the card). While most banks attacked both fronts, the signing of merchants to accept the card generally received more emphasis. Therefore, as consumers were solicited to carry the cards, they saw more and more places to use them. This accounts for the early success of the card—it was highly convenient to use. As more merchants signed on, the numbers of cardholders grew. Banks promoted the convenience factor heavily: no problems with checks, no need to carry large amounts of cash, no difficulty purchasing goods or services, and wide acceptance by many merchants.

In addition, the bankcard made credit readily available. For consumers who could not afford or did not want to make a purchase with their own funds, credit from the issuing bank was now available through the card. The amount of credit used could be paid in full by the payment due date, or the amount borrowed could be repaid in flexible monthly installments. For the first time, bank customers could use credit without having to go to the bank to apply for a loan. The card provided credit to the customer, even when the customer was far from home. Convenience and credit availability were major contributors to the early proliferation of credit cards around the country.

Another feature that made the card attractive to consumers was that payment on purchases could be delayed for approximately one month. Even if the customer had sufficient cash to pay the balance in full, the monthly statement was not received for some time after the purchase was made. Today, more than 70 percent of all cardholders do not pay the outstanding balance in full.

country club billing
A billing system in which copies of transactions are mailed to the cardholder with a monthly statement.

Recordkeeping was facilitated as well. At the time of each purchase, the cardholder was given a receipt. When the statement came from the bank, the original draft or sales slip from the merchant was included. Matching the receipt with the merchant draft confirmed that the cardholder made the purchase and helped reconcile the balance owed to the bank. This was called country club billing. Now, of course, transactions are printed on the statement in chronological order and the merchant draft is not returned to the cardholder—a practice called descriptive billing.

descriptive billing
A method of billing in which each monetary transaction posted to an account during a billing period is identified and described on the bill.

The net result was that consumers found credit cards easy to use. Credit was now immediately available to fund everyday transactions when the cardholder wanted to avoid using personal funds. The accumulated transactions were easy to account for each month. And the number of merchant locations at which the card was accepted was growing steadily.

Merchants

From the merchant's point of view, the card had several attractive features. First, the sales transaction was fairly easy to validate. The procedure was, and continues to be, simple. If the account number did not appear on the merchant warning bulletin and was under the floor limit, the cardholder's purchase was considered valid. If the purchase exceeded the floor limit, the merchant simply called the bank's authorization center to have the sale validated.

The merchant had none of the risks inherent in extending credit or accepting checks. There was no need for the merchant to contact a purchaser directly in an attempt to recover money. By following the authorization procedure, the sale was validated, and the merchant would receive payment from the bank upon deposit of the signed draft.

A second reason merchants liked the card was the opportunity to increase sales. Banks and the national associations (MasterCard International and Visa International) spent large sums of money on advertising. The advertisements featured decals on display in the windows of merchants that accepted the cards. The likelihood that customers would patronize these merchants increased dramatically as more cards were issued. This was especially true of merchants that had previously accepted only cards under their own store name. Many merchants showed an early increase in sales that was attributable to the card. The average credit card transaction was larger, and consumers bought more with their credit cards than with cash.

Handling credit purchases was virtually impossible for smaller merchants before the advent of the bank credit card—a third reason for its appeal. A merchant credit plan was difficult to establish and expensive to administer. Merchants without credit plans could not make sales to consumers who wanted to make a purchase on credit. MasterCard International and Visa International changed the rules of the game.

Any merchant that accepted the card could make a sale to any consumer who carried the card. No consideration of the consumer's creditworthiness was required. What appealed to smaller businesses, in many cases, also appealed to medium-sized and large businesses. Although some major department stores chose not to accept bankcards in the early days, many others did, and they enjoyed the benefit of being able to make sales to an ever-expanding universe of cardholders.

Easy validation of sales, expanded sales opportunities, and fewer credit plans played a key role in merchants' acceptance of the cards. However, some merchants decided not to accept cards. The discount rate was found to be the primary reason in these instances.

Banks

There were some important advantages to banks that offered cards to their customers. In the early days, the revolving line of credit associated with the card was probably the main attraction. Under a revolving line of credit, the customer could borrow money, repay the amount, and borrow again without having to go into the bank to apply for another loan. As long as the total amount borrowed did not exceed the credit limit—the total dollar amount of credit the bank had made available to the customer—the customer could borrow repeatedly and repay the amount owed in monthly installments. Banks used the card as an easy way to extend credit and as an incentive for customers to borrow. Interest income grew as more customers borrowed money and repaid only a part of the balance due each month. Some of the problems created by this easy access to credit are discussed in chapter 2.

A second advantage the card gave to banks was that it attracted customers who did not live nearby. Traditionally, customers would choose a bank because it was convenient to their home or job. With a card, the bank's location was irrelevant. Before long, banks began to use this remote feature to expand their market areas and offered their cards to consumers who lived hundreds or thousands of miles away. Unlike conventional lending relationships in which the customer came into the bank every time a loan was needed, credit could be extended without inconvenience to the borrower. Banks saw their outstanding loans grow quickly as a result of card programs.

Banks also viewed new cardholders as prospects for additional bank products. Because new customers were being attracted from geographic areas not previously within their reach, new opportunities arose to establish additional relationships. In fact, many banks heavily engaged in attempts to cross-sell additional banking products and services to cardholders.

While consumers offered new sources of income to banks as a result of the card, so did merchants: merchants paid discounts to banks for handling cardholders' purchases. Although pricing in the early days was not highly sophisticated, it was competitive. Banks found a new reason to market aggressively to merchants. An aggressive marketer could leverage the merchant relationship into sales of other products.

In addition to discount income, merchants brought additional deposits to the bank. The merchant sales draft was treated as a deposit item. Therefore, dollars generated by consumer purchases made with the card generated new deposits for the bank, which in turn became new sources of funds for making loans.

With the rapid growth of credit cards, banks actively sought to become the merchant bank. The combined attraction of new income and new sources of deposits created a high level of competitive activity. As more merchants accepted the card, the convenience to cardholders expanded. With more cardholders, merchants enjoyed new sales opportunities. Each reinforced the other and spurred the growth in the early years. Meanwhile, banks enjoyed new benefits from both cardholders and merchants.

Growth Pains

The virtual explosion of bankcards in the early years was not without problems. The banking industry had never before experienced the unique growth dynamics of credit cards, and inexperience resulted in some adverse consequences.

A Learning Experience—Revolving vs. Installment Credit

When bankcard programs were first established, no state or federal law prohibited cards from being issued on an unsolicited basis. Banks were free to send cards to any person considered worthy, whether or not the recipient had requested it. This advantage caused banks to engage in the mass issuance of cards to existing and prospective customers. Credit lines were established and cards mailed. Banks sent cards to deposit customers, loan customers, safe deposit customers, and any other customers whose addresses they could obtain. Many banks bought mailing lists consisting of names from magazine subscriptions, driver's license registrations, and the like. Not surprisingly, some of the individuals who were issued cards did not manage the credit well. Credit losses began to mount, and in just a few years, the losses soared.

Banks generally had no experience in extending unsecured credit to large numbers of people—especially revolving credit associated with a card. The credit approval criteria that had served banks well when making installment loans, such as automobile or house loans, proved to be inadequate for extending credit through cards. New approval criteria had to be developed. Unfortunately, this did not happen until after banks had incurred unacceptable losses.

In virtually all cases, the revolving credit line was unsecured: no assets of the borrower were available to be recovered and sold to offset a portion of the loss. Once known to be a loss, the entire amount had to be charged off. In a significant percentage of cases, the credit line had been exceeded before the card could be recovered or blocked, so the amount charged off was greater than the amount of credit originally extended to the cardholder.

charge-off
(1) The balance on a cardholder account that a bank no longer expects to be repaid and writes off as a bad debt. (2) The process of charging off accounts, which is generally recorded by a debit to the reserve for possible credit losses and a credit to the loan balance. Also called bad debt.

The loss problems were aggravated in the beginning because authorization procedures were cumbersome and slow. For example, putting an account number on the merchant warning bulletin could take several weeks. As long as the delinquent customer kept the amount of purchase below the floor limit, the merchant had no way of knowing that the cardholder's charge privileges had been suspended. Delinquent borrowers could make numerous purchases of this nature before the account appeared on the bulletin.

For all of these reasons, bright prospects dimmed as credit losses climbed. Banks eager to issue cards to as many people as possible began to pull back as growing bankcard portfolios began to turn sour. Both consumers and banks had to learn some hard lessons concerning the proper handling of bankcard credit.

Fraud Losses

So many cards were issued, and so many merchants accepted them, that it is not surprising that criminal activity became involved. An underground network was established to facilitate the illegal use of cards. For example, a card stolen in California on day 1 would show up in New York or Florida on day 2. As a result, fraudulent transactions began to mount. Over time, criminals began to counterfeit and alter cards. Even dishonest merchants found ways to defraud the banks by creating false drafts on real cardholder account numbers, depositing the drafts in the bank, and withdrawing the cash—a practice known as white plastic fraud.

Cumbersome authorization systems hampered the ability of banks to stop the counterfeiting, altering, or stealing of cards. Fraud is now considered one of the costs of doing business. While fraud accounts for the loss of millions of dollars annually, it does not have a significant impact on bankcard profitability. (See chapter 10 for an in-depth discussion of bankcard fraud.)

Deregulation

During the early days of bankcards, strict regulations concerning usury ceilings, credit limits, and the ability to assess annual membership fees conspired to make it very difficult for individual bankcard programs to operate at a profit. It is easy to understand why profits were scarce when one understands that for several years, the spread on bankcard outstandings was very small and sometimes negative.

usury
(1) A higher rate of interest than allowed by law.
(2) The act of charging a higher rate of interest for the use of funds than is legally allowed by a state.

spread
The difference between the bank's cost of funds and its interest yield. Example: If the cost of funds is 10% and the yield on those funds is 15%, the spread is 5%.

In the early 1980s, several states began to loosen their regulatory constraints. The first two states to deregulate were Delaware and South Dakota. Deregulation meant that these states allowed issuers who based their operations there to charge interest rates without usury ceilings; to charge those rates throughout the rest of the country; and to charge fees, such as annual membership, late fees, overlimit fees, etc. Banks migrated to these deregulated states to increase their profitability. The states that deregulated did so to attract major bankcard organizations to their state. Their efforts, and the efforts of states like Georgia and Nebraska, proved to be highly successful. By now, virtually all the major issuers have relocated to a handful of states with liberal regulations.

Nonbank Entry

From their beginnings in the early 1960s until the mid-1980s, bankcards were the exclusive domain of the banking industry. By that time, bankcards had progressed from "loss leader" at best to being one of banking's most profitable services. Thus, it is not surprising that nonbank companies were attracted to the business.

When Sears launched its Discover card in 1985, conventional thinking within the banking community suggested there was little chance of success. This was partially because of the belief that the bankcard market was saturated. By the end of 1990, Discover was the third largest card issuer in the United States. Today, it is reported to be profitable and to have recovered its start-up costs, which included the considerable expense of establishing an independent merchant network. Finance companies have also made their presence felt. The Associates, Household, and Beneficial have become major players in the industry.

Nonfinancial entities are having a significant impact on the bankcard business, either by partnering with a bank issuer or buying/chartering their own credit card bank. These organizations cover a wide range of industries—communications, mutual funds, insurance, transportation, etc.—but they have several points in common. They all have a large retail customer base and operate in a highly competitive industry. Their principal focus has been on maintaining and enhancing their core business rather than maximizing the profitability of the card. The most notable of these issuers is AT&T, which in 1990, its first year, opened more than five million accounts.

Summary

The relatively brief history of bankcards has been dramatic in terms of growth and change in the banking industry. John C. Biggins of Flatbush National Bank in New York was the first to implement the idea of bankcards. These cards became a matter of national significance in the 1960s with the introduction of BankAmericard (now Visa International) and Interbank Card Association (now MasterCard International). The synergies of the two national card systems strengthened the dramatic growth of bank credit cards in the years that followed.

The major reasons behind rapid growth must be considered from the perspectives of the consumer, the merchant, and the bank. For the consumer, the bankcard made purchases of products and services more convenient, especially when credit was desired to fund these purchases. Bank customers could obtain credit for a variety of purchases without repeatedly going to the bank for a loan. The amount owed could be paid in full each month or extended through monthly installments.

The merchant found the bankcard attractive because sales transactions could be validated easily and payment guaranteed. Heavy promotion of the card by banks and the national associations increased the sales opportunities for merchants who accepted the cards. The associations likewise relieved merchants of the risk and cost of in-house credit plans.

Banks found an attractive way to extend credit to consumers through the revolving line of credit attached to the bankcard. Geographic market areas were expanded because banks could issue cards to customers who did not reside near the bank. With these new customers came additional opportunities to sell other banking products. Income from cardholders was complemented with new income sources from merchant discounts and new deposits from sales drafts.

Rapid growth in credit brought new levels of credit losses. In many cases, approval criteria were inadequate for numerous lines of unsecured revolving credit. Early authorization systems were slow and use of the card was difficult to curtail. Fraud losses also aggravated the problem. Lost or stolen cards, altered or counterfeit cards, and white plastic merchant fraud all rose to a level of national concern in the early 1980s. Management frequently had difficulty coping with the dynamics and magnitude of bankcard problems. Nevertheless, these programs survived.

Review Questions

1. Name five types of bankcards and discuss their distinguishing characteristics.

2. From the consumer's point of view, what are some reasons for the early growth of bank credit cards?

3. Why was the bankcard attractive to merchants in the early days?

4. What is the difference between descriptive and country club billing practices?

5. What role did revolving credit play in the growth of the bankcard business? Were there any adverse effects?

2

BANKCARD ORGANIZATION

This chapter outlines the functional roles played by the national associations, as well as the conventional organization structure established by bankcard issuers and acquirers to operate and manage the bankcard business. You will learn the functions involved in operating a bankcard program and the departmental areas normally responsible for their execution.

After successfully completing this chapter, you will be able to

♦ identify the national bankcard associations and discuss their general operation

♦ outline the functions of the processing systems of the national bankcard associations

♦ explain the functions of a bankcard issuer

♦ define the terms duality, interchange, settlement, issuer, and acquirer

♦ list and explain the options available to a bank in structuring its bankcard operation

National Associations

Both Visa International and MasterCard International, the two national associations, played critical roles in the early and ongoing success of the major bankcard programs. Both are owned by member banks and governed by separate boards of directors. Funded through assessments and fees charged to member banks based largely on bankcard volumes, the national associations perform several key functions:

◊ licensing
◊ patents and copyrights
◊ operating regulations
◊ national and international settlement and authorization systems
◊ interchange
◊ research and analysis
◊ product development
◊ advertising and promotion

The MasterCard and Visa organizations have kept the bankcard industry on the leading edge of technology, particularly through networks and automation. By providing a means to bring together the collective resources of banks and other financial institutions, the associations have accomplished what no member could have achieved independently.

Major Operating Systems

Each association owns and operates international processing systems that provide capabilities to authorize purchases and settle merchant and cardholder transactions in the United States and abroad. For example, when a cardholder from Texas wants to buy an item in Paris, France, the authorization to validate the purchase for the Paris merchant is obtained through a system run by one of the associations.

authorization
Approval by, or on behalf of, the card-issuing bank that validates a transaction for a merchant or another affiliate bank.

After the purchase has been completed, the merchant must be paid in French francs and the cardholder billed in U.S. dollars. The bulk data transfer systems that process such transactions, including currency conversion, are Banknet for MasterCard International and Visanet for Visa International. These authorization and settlement systems handle billions of transactions annually.

Interchange

A critical function managed by the national associations is that of interchange. As the word suggests, this function enables banks around the world to exchange information, transactions, money, and other items on a standardized and consistent basis. Without standardization, the bankcard industry would collapse. For example, through the association network, the authorization system in Paris, France, recognizes the card number format the same way the merchant recognizes the graphic on the face of the card. For electronic interchange to take place, standardization of minute details is necessary, such as the exact positioning of embossed characters on the card or data elements in the magnetic stripe. This allows banks around the world to communicate with each other.

Interchange Fee

An important component of the interchange function is the fee. The purpose of the interchange fee is to compensate the cardholder bank for the "free" period between settlement, or payment, to the merchant bank for cardholder purchases and billing to cardholders. Other operating costs incurred by the cardholder bank are also considered in the interchange fee. The national associations set and regularly adjust the fee. There is one interchange fee for MasterCard International transactions and another for Visa International transactions.

The flow of fees is shown in exhibit 2.1. You can see from this example that if the purchase amount is $100 and the interchange fee is 1.5 percent per transaction, the amount of the interchange fee is $1.50.

EXHIBIT 2.1 Transaction Flow and Interchange Fees

The merchant bank pays the interchange fee to the cardholder bank through the settlement system. The dollar value the merchant bank receives for merchant transactions sold to interchange is discounted to the merchant bank. The fee is income to the cardholder bank but an expense to the merchant bank. We will discuss this further under profitability in chapter 3.

Cash advances are the exception to the flow of the interchange fee. In the case of a cash advance, the cardholder bank reimburses (through interchange) the bank that made the cash advance to the cardholder.

Different interchange fees are applied depending on whether or not credit card transactions are electronically authorized at POS terminals and entered into interchange for prompt settlement (within a specific number of days after the purchase date). The fees are set by the national associations and adjusted periodically. Though intended for the same purpose, the electronic-based fees are called by different names: AID (Acquirer Interchange Discount) for MasterCard and EIRF (Electronic Interchange Reimbursement Fee) for Visa. Therefore, interchange refers to standardization for authorization and transaction settlement, as well as to the fee paid by merchant banks to cardholder banks.

Settlement

Settlement is the process by which the dollar amounts of cardholder purchases are passed from the merchant bank to the cardholder bank. In the case of international transactions, the currency involved in the cardholder purchase is converted to the national currency appropriate for billing to the cardholder. For example, a purchase made in France using French francs would appear on an American cardholder's statement in U.S. dollars. Through interchange, the merchant bank "sells" cardholder purchase amounts to the cardholder bank, which "buys" the cardholder purchase amounts and subsequently bills the cardholders. The purchase is sold by the merchant bank through interchange to the cardholder bank for billing to the cardholder.

Net settlement more accurately describes the process used. The net difference between cardholder purchases to be sold (by the merchant bank) and the cardholder purchases to be bought (by the cardholder bank) is the actual amount settled in interchange. Let's assume a bank has both merchants and cardholders. At the end of the day, all the merchants have deposited $100 in drafts (cardholder purchases). The bank will sell the total of $100 through interchange to other cardholder banks. However, at the end of the same day, the bank's cardholder purchases of $75 are coming in through interchange to be bought by the bank. The interchange system nets the amounts for settlement, as follows:

Merchant deposits (outgoing interchange)	$100 Sells
Cardholder purchases (incoming interchange)	$-75 Buys
Net settlement the interchange system paid to bank	$ 25 Settles

Duality

One of the controversial changes in the bankcard arena came in 1976 with the change of regulations that allowed banks to become members of both MasterCard International and Visa International. Before 1976, a bank could be a member of only one association or the other, but not both. As a result of the assessments required by both associations, bankers began to debate the question of whether two associations were needed, as indicated in the following excerpt from *Bankers' Desk Reference: 1985.*

> The two national associations . . . have experienced unrest in recent years, to a point where the continuing role of the two associations is being challenged. With the advent of duality in the 1970s came questions about the *need* for two organizations with some redundancies of function and duplicate costs. In the 1980s, the questions have become issues. At the 1983 National Bankcard Convention in Los Angeles, for example, the possibility of merging processing operations was discussed publicly for the first time. Subsequent discussions have taken place, and there is some expectation that a number of operating functions will be merged. [Some have been since 1983.]

This debate has settled somewhat since 1985. The general view is that both associations will continue. But the consequences of duality will not be soon forgotten. The competitive battle for merchants (through pricing competition) has reduced profitability and increased the risk of credit problems. As a result of duality, cardholders were frequently offered twice the amount of credit as was previously extended. More overextensions and losses resulted. Duality changed the nature of bankcard competition. For better or worse, the changes brought about by duality will be felt for years to come.

Issuers and Acquirers

During the initial growth of the industry, and until the 1980s, it was normal for banks to be both issuers and acquirers. During the late 1980s, the number of acquirers was substantially reduced to the point where a small number of major banks, together with a few nonbank merchant processors, such as Nabanco and National Data Corp., processed the lion's share of merchant transactions.

Issuer Organization

In bankcard terminology, the issuer is concerned with all aspects of card issuance and the cardholder relationship. The card-issuing business can be divided into 12 functions:

♦ **Administration**. Responsibility for the overall management of the business is vested in the center administration.

♦ **Marketing**. This unit is responsible for developing and implementing cardholder solicitation and activation programs.

♦ **Credit Processing**. This unit is responsible for the operational activities connected with the credit decision and set-up of new accounts.

♦ **Card Issuance**. This department embosses, encodes, and issues plastic cards to new customers and reissues cards to existing customers.

♦ **Incoming Interchange**. This area processes incoming cardholder transactions and cardholder chargebacks.

♦ **Cardholder Billing**. This group prepares and mails cardholder billing statements, as well as other system-generated cardholder correspondence (for instance, late notices).

♦ **Payment Processing**. Activities associated with processing and posting cardholder remittances are performed by this unit.

♦ **Customer Service**. All cardholder inquiries, complaints, and requests— except those concerned with credit, collections, or fraud issues, are handled by the customer service department.

♦ **Overlimit**. This is generally a relatively minor function, and it may be handled either in the collection or credit unit. Its purpose is to monitor and cure situations where a cardholder's balance exceeds the credit limit assignment.

♦ **Collections**. This area is responsible for collecting outstanding balances from delinquent and charged off accounts.

♦ **Fraud Control**. This department is responsible for handling reports of lost or stolen cards and reviewing activity on blocked accounts.

♦ **Cardholder Authorization**. This is generally a highly automated function, which is responsible for updating the cardholder masterfile with authorization requests, and for liaison with the national associations, which process the majority of authorization requests.

Exhibit 2.2 depicts the conventional structure of a card-issuing business organization.

EXHIBIT 2.2   Bankcard Issuer Organizational Structure

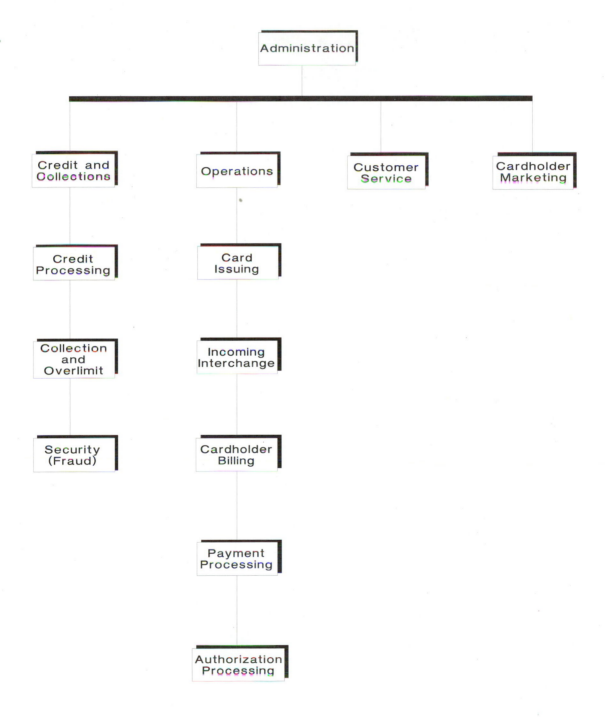

Acquirer Organization

In bankcard terminology, the acquirer is concerned with all aspects of the merchant side of the business, which can be divided into six functional areas:

♦ **Center Administration**. This area is responsible for the overall administration and control of the merchant relationship.

♦ **Proof and Capture**. This area processes and reconciles all bankcard transactions made at the acquirer's merchant outlets and deposited to the merchant's account.

♦ **Outgoing Interchange**. This area processes outgoing sales drafts and cash disbursements upon completion of proof and capture. Merchant chargebacks and data retrieval requests are also handled in this area.

♦ **Risk Management**. This unit processes new merchant applications and assures the creditworthiness of new and existing merchants. It also monitors merchants for suspected fraudulent activity.

♦ **Merchant Sales**. This department solicits and signs up new merchants and provides both in-house and field support to the merchant base.

♦ **Merchant Authorization**. This function provides a timely response to electronic and voice authorization requests from member merchants.

Bank Options

The functional responsibilities of bankcard issuers are well defined and vary little from bank to bank. However, each bank can choose from several alternatives in deciding how to deliver its services.

♦ **Standalone**. This is the traditional in-house processing method, in which the bankcard operation is an independent, self-contained business unit and does all its own processing. This is the option most often chosen by the very largest card issuers.

♦ **Regional Associations**. Regional bankcard associations were originally founded and owned by groups of banks in a regional area. Their purpose was to provide an economy of scale that none of their members could attain on a standalone basis. These associations generally performed back-end processing for their member banks, while each bank member provided its own front-end servicing.

back-end processing
Generally refers to activities that do not involve customer contact or risk management. Authorization and cardholder billing are considered part of the back-end processing, while activities like credit review and customer service are considered front-end services.

The 1980s saw both a dramatic change in the structure of regional associations as well as a reduction in their number. In some cases, the association was spun off—sold—by its member bank owners and became a for-profit corporation. An example of this is the Mid-America Bankcard Association (MABA), which was founded by several banks in Nebraska and nearby states. After several years, the name was changed to First Data Resources (FDR), and its services were expanded and marketed to nonmembers on a fee basis. The company was eventually purchased by American Express and, by the end of 1990, FDR was the largest third-party processor of bankcards in the country, servicing in excess of 40 million accounts. Other regional associations have either been purchased and absorbed by a larger association or become a captive in-house processor for a major issuer. The purchase of Eastern States Bankcard Association (ESBA) by FDR is an example of the former, while MBNA's buyout of Southern States Bankcard Association (SSBA) is a example of the latter.

♦ **Outside Service Providers**. Another option available to bankcard issuers is to employ an outside service company to perform one or more of the functions involved in operating a bankcard program. Outsourcing certain operations is frequently a cost-effective alternative, particularly for smaller issuers. The kinds of services most generally purchased are

◊ data processing
◊ authorization
◊ card production
◊ marketing

outsourcing
The use of outside vendors to perform technical, behind-the-scenes functions.

Since the late 1980s, the trend has been toward increased outsourcing. This trend coincided with the pressure on a bank's overall profits caused by problems elsewhere in the bank—notably in real estate loans and loans to third world countries—and allows small banks to be part of the bankcard business without major capital investment.

Summary

This chapter has looked at the functional responsibilities that comprise the operation and management of a bankcard program. The national associations have played pivotal roles in facilitating the movement of transactions between the merchant, the acquirer, and the issuer in an orderly and systematic fashion. The functions performed by the issuer (the cardholder bank) and the acquirer (the merchant bank) are conventionally organized. Banks have several alternatives in deciding how best to provide their services: for large banks, a self-contained in-house unit is often the most viable; regional associations formed by a consortium of geographically linked banks is another alternative; and, increasingly, many banks are choosing outsourcing to provide front-end and/or back-end support.

Review Questions

1. How many national associations are there? What are their names? Who owns them?

2. Name two functions performed by the processing systems owned by each national association.

3. Name and explain 3 of the 12 functions performed by a bankcard issuer.

4. Define acquirer. Is the trend toward more or fewer acquirers?

5. Name and briefly explain the three options available to a bank in structuring its bankcard organization.

3

BANKCARD PROFITABILITY

So far, we have reviewed the history of the bankcard industry and learned how banks typically organize their bankcard departments. This chapter discusses the profitability of bankcards. You will learn which elements affect bankcard profitability and gain an understanding of the profit dynamics of the bankcard business from the perspective of both issuing and acquiring banks.

After successfully completing this chapter, you will be able to

♦ discuss the rationale on which bankcard profitability was originally based

♦ identify the elements that determine the profitability of an issuing bank's program

♦ discuss the outside factors that exert pressure on a program's profitability

♦ define return on assets and return on expenses

♦ identify the elements that affect the acquiring institution's profitability

Evolution of Bankcard Profitability

The original concept for operating a bankcard program was based on the assumption that approximately two-thirds of the profits would come from cardholders, while the merchant side of the business would generate one-third of the profits. In practice, merchant-generated profits fell short of expectations from the outset. In the first place, merchants were quick to realize that the success of this business depended on the merchants' willingness to accept the cards as a method of payment. Second, merchants were able to take advantage of the fierce competition among the banks to sign them up. These factors had the net effect of forcing the merchant discount rate down. Finally, many of the costs associated with operating the merchant side of the business, such as authorization and fraud, were higher than had been anticipated.

As the proliferation of bankcards gained momentum in the 1960s, many banks adopted a defensive marketing strategy. They felt obliged to include a bankcard among their retail services to prevent losing their existing base of consumer deposit and loan customers to competitors who were actively promoting the card. Other factors also had an adverse impact on cardholder profitability in the early years:

- ◊ mass marketing of unsolicited credit cards
- ◊ inexperience in evaluating unsecured, revolving credit risk
- ◊ impractical regulatory restrictions, primarily usury ceilings
- ◊ cumbersome, labor-intensive systems and procedures

Over the past several years, more sophisticated marketing, credit evaluation, and system support have more than offset early setbacks. Banks have learned the business and taken the steps necessary to permit them to operate their programs profitably. During the 1980s, bankcards emerged as one of the most profitable services offered by banks.

Measurements of Profitability

Issuer Economics

Profit or loss is the difference between revenue and expense. In addition to measuring profits in absolute terms of dollars earned, banks measure the profitability of their cardholder portfolios in terms of their return on assets (ROA).

Bankcard ROA is defined as net earnings before taxes expressed as a percentage of the dollars outstanding. For example, if Bank A has a bankcard portfolio with $1 million outstanding and earns $35,000, the ROA is 3.5 percent ($35,000/$1,000,000 = 3.5%).

Using ROA as a standard of measurement enables banks to compare their relative performance. To demonstrate, let us assume that Bank A earns $35,000 on outstandings of $1 million, and Bank B earns $100,000 on outstandings of $5 million. It is clear that Bank B has earned more money. However, you can also see that Bank A has made better use of its assets, as shown below:

Bank A $35,000/$1,000,000 = 3.5% ROA
Bank B $100,000/$5,000,000 = 2.0% ROA

Elements of Profitability

Nine basic elements of revenue and expense determine the profitability of a bankcard-issuing program:

- ♦ **Interest income** is determined by multiplying the annual percentage rate (APR) by the rollover rate by the outstanding dollars.

rollover rate
The percentage of dollars outstanding that are subject to finance charge. Example: If a portfolio has $1 million outstanding and $700,000 is subject to finance charge, the rollover rate is 70% ($700,000/$1 million).

- ♦ **Interchange income** is the percentage of each dollar of merchant sales that the acquirer is required to pay the issuer.

- ♦ **Annual membership fee** (AMF) is the fee the cardholder pays to the issuer for the right to carry and use the bankcard.

- ♦ **Exception processing fees** are miscellaneous charges levied against a cardholder for not following the rules of the cardholder agreement. These charges are often called nuisance fees and are levied for such infractions as late payments, exceeding one's credit line, and having a check returned.

- ♦ **Cost of funds** is the largest single expense item associated with a card program. It is defined as the interest rate a card issuer pays to finance the outstanding balances in the bankcard portfolio.

- ♦ **Losses** include bad debts and fraud losses and are usually expressed as a percentage of outstanding balances.

♦ **Servicing expenses** refer to front-end costs (discussed in chapter 2) and include risk management and customer service activities.

♦ **Processing expenses** represent the sum of the back-end costs as outlined in chapter 2.

♦ **Marketing expenses** include all costs associated with acquiring new accounts, increasing usage on existing accounts, and retaining profitable accounts.

Each of the nine elements of profitability discussed above is composed of many subelements. For example, cost of funds (COF) is a composite of the costs a bank incurs to obtain demand deposits, savings deposits, certificates of deposit (CDs), etc. As a general rule of thumb, a bank's COF can be assumed to be 1½ to 3 percent below the prime rate.

prime rate
The rate a bank charges its best commercial customer to borrow funds.

Profit and Loss Modeling

We have looked at the elements of bankcard profitability. Now we need the following information to construct a profit and loss model for a bankcard issuer program:

◊ total dollars outstanding
◊ number of cardholder accounts
◊ average balance per cardholder account
◊ annual percentage rate
◊ rollover rate
◊ dollar sales
◊ interchange rate
◊ annual membership fee
◊ exception processing fees
◊ cost of funds rate
◊ credit and fraud losses
◊ servicing costs
◊ processing expenses

Exhibit 3.1 shows the format for developing a profit and loss model for a card-issuing bank.

EXHIBIT 3.1 · Bankcard Issuer Profit and Loss Model (active accounts)

	Average outstanding $
	Per account _____
	Percentage of outstandings

Revenue
  Interest income (% APR x % rollover rate = interest yield)
  Interchange income ($ sales x interchange rate/average balance = yield)
  Annual membership fee ($ AMF/$ average balance = yield)
  Exception processing fees (total fees/average balance = yield)

                                                  Total Revenue:

Expense
  Cost of funds
  Losses (total losses/average balance = % of outstanding balance)
  Servicing (servicing expenses/average balance = % of outstanding balance)
  Processing (processing expenses/average balance = % of outstanding
     balance)
  Marketing (marketing expenses/average balance = % of outstanding
     balance)

                                                  Total expense:
                              Net earnings before taxes (NEBT)/ROA:
                                 Pretax profit per active account:

Acquirer Economics

The merchant or acquiring side of the credit card business has never achieved the revenue targets that were projected when the business was launched. Fierce competition for merchant business has resulted in lower-than-expected merchant discounts. As a result, the business has become polarized, so that today there are relatively few players who are both acquirers and issuers. As with so many other businesses, economies of scale have enabled large acquiring institutions to achieve profitability. Six basic elements influence merchant profitability:

♦ **Merchant discount** is the fee the acquirer charges the merchant in return for buying bankcard sales generated at the merchant location. It is the product of the average discount rate times the bankcard sales volume. For example, if a merchant is assigned a discount rate of 2.2 percent and generates $1 million in bankcard sales, the merchant discount is $22,000.

- **Other income** is revenue generated primarily from the sale or rental of such items as POS terminals and imprinters.

- **Earnings on deposits** comes from net interest earned on merchant checking account deposits.

- **Outgoing interchange** is the largest expense. It is a percentage of each merchant sales dollar which the acquirer is required to pay the issuer.

- **Processing expense** is the acquirer's cost of supporting the merchant. It includes expenses associated with authorization, merchant accounting, supplies, etc.

- **Sales and marketing expenses** include the cost of the merchant sales and support staff and services.

The combined average experience of Visa and MasterCard acquirers for 1990 provides an example of a profit and loss model for acquirers (exhibit 3.2).

EXHIBIT 3.2   Acquirer Profit and Loss Model

Revenue

Merchant sales	$1,000,000
(average ticket = $70)	
(average discount rate = 2.1%)	
Gross discount income	21,000
Other income at 5% of discount	
(imprinter/terminal rental)	1,050
Net earnings on deposits at 5 basis	
points of sales	500
Gross revenue:	$22,550

Expense

Outgoing interchange at 1.4% of sales	14,000
Processing expense at $.41 per ticket	5,857
Sales/marketing expense	500
Credit and fraud loss at $.022 per ticket	315
Total expense:	$20,672

Net earnings before taxes:	$1,878
Return on sales:	18.8BP
Return per ticket:	$0.131
Return on expenses:	28.1%

From the model, it is clear that an acquirer's revenue comes primarily from the merchant discount. While technically a variable element, in recent years the range of discount rates for similar types of merchants has become much narrower. Large national and international merchant chains, such as airlines, hotels, and car rental companies, have come to understand the economics of the acquiring function and have successfully negotiated discount rates that effectively control the acquirer's principal revenue source. Acquirers earn other revenue from the sale or rental of POS terminals and imprinters, but this is a relatively small amount. Net earnings on merchant deposits—the difference between what the acquirer earns on deposits and what it passes back to the merchant—is a third source of acquirer revenue.

Outgoing interchange is the largest expense incurred by the acquirer. It is periodically established by the national associations and is meant to reimburse the card issuer for its costs associated with the transaction. In 1992, the effective rate for outgoing interchange is approximately 1.5 percent of sales.

The next biggest expense incurred by an acquirer is the cost of processing merchant sales transactions. Based on a 1989 cost study performed by Visa International, the average cost per ticket was $0.41. Clearly, this area represents the acquirers' principal opportunity to control profitability. Economies of scale are critical, and this factor is largely responsible for the polarization of issuing and acquiring banks. In exhibit 3.2, for example, if an acquirer were able to reduce processing costs by 5 percent (from $0.41 per ticket to $0.39 per ticket), profit would rise by 15 percent (from $1,878 to $2,164). The sales and marketing expense item is variable, but it is generally not a major factor in the equation. Credit and fraud losses are unpredictable and depend on the acquirer's diligence in qualifying each merchant and monitoring merchant performance. The 1989 Visa cost study found these losses to average $0.022 per ticket.

Profit (or loss) on the acquiring side can be measured in four different ways:

♦ **Net Earnings before Taxes.** This is the dollar difference between revenue and expense. It is the only absolute measure of profitability, but it does not indicate the efficiency of the operation.

♦ **Return on Sales.** This is generally shown in basis points and provides a method of comparison between acquirers of different size. It is determined by dividing net earnings before taxes by total merchant sales and is normally expressed in basis points. (Each basis point is equal to 1/100 of a percent.)

♦ **Return per Ticket.** This is the best indication of an acquirer's efficiency since it measures profit per transaction. It is calculated by dividing net earnings before taxes by sales transactions.

♦ **Return on Expenses.** This is calculated by dividing net earnings before taxes by variable expenses (that is, all expenses except outgoing interchange). This is the most common measurement of an acquiring business's profitability.

Profit Pressures

The number of bankcards outstanding in the United States has grown from about 50 million in 1970 to more than 200 million in 1990. Because so many people now hold at least one bankcard, it is easy to understand why competition for new cardholders has become so keen.

However, funding became a problem as the interest rates banks paid to fund credit card portfolios took dramatic swings in the 1970s and 1980s. New funding strategies, such as issuing commercial paper and securitizing bankcard receivables, were developed, and new pricing structures were imposed. Although the number of cards continues to grow, the rate of growth slowed in the late 1980s. This slower growth is reflected in reduced response rates to direct mail and telemarketing programs. The lower response results in a higher marketing cost for each new account, which in turn affects profits. Chapter 4 discusses the role and impact of marketing in greater detail.

commercial paper
Negotiable, short-term, unsecured securities issued by major businesses to raise money.

securitization
Bundling of loans (e.g., credit card receivables) into packages to sell as investments secured by those receivables.

Increased competition for new accounts has caused issuers to attempt to differentiate their card from others by adding value to their program. This added value may take the form of lower charges to the cardholder or additional features, such as extended warranty on purchases. However, product differentiation puts pressure on a bankcard program's profit.

Risk management is the term used to cover the credit, collection, and fraud functions of a bankcard program. How well an institution performs these functions has a direct bearing on the bottom line. In later chapters, we will discuss these functions and some of the techniques and tools which are available to the risk manager.

Summary

The bankcard business has evolved from a loss leader in its infancy to its present status as the most profitable segment of many banks' businesses. This chapter has looked at the elements that determine profitability and how these elements affect the bottom line. Increasing competition for cardholders, and the efforts of bankcard issuers to differentiate their cards by adding enhancements, has put pressure on the profitability of bankcard programs. From the merchant side of the business, profitability can be achieved by economies of scale and is measured by the return on expenses.

Review Questions

1. When the bankcard industry began, what portion of the total revenue did the founders expect to come from the cardholder? From the merchant side of the business?

2. Name and explain the terms by which bankcard profitability is measured and compared.

3. What is the largest expense element that is totally within the control of the acquirer?

4. Has competition had an effect on profitability? How?

4

MARKETING

The marketing of bankcards has a great deal in common with the marketing of other financial products. Both the level and the nature of the competition for bankcards have brought about some important differences. On the cardholder side, the attractive yield on the asset (ROA) has drawn the attention of banks, thrift institutions, finance companies, and other consumer financial service organizations. The eighties saw a period of fierce competition for cardholders. The ability of financial institutions to reach beyond traditional geographic boundaries by issuing bankcards heightened competitive interest. Each new bankcard customer also brought the potential for additional financial relationships, including the opportunity to sell consumer merchandise and nonbank services such as travel packages. Bankcards opened new markets that in turn attracted new competitors to the banking arena.

After successfully completing this chapter, you will be able to

♦ discuss the role of marketing in establishing and maintaining a bankcard program

♦ cite the major elements involved in developing and executing marketing strategy as it relates to a bankcard program

♦ explain the importance of marketing techniques in improving overall bankcard profitability

♦ define attrition, grace period, and skip payment

Account Acquisition

The livelihood of a bankcard program is new accounts. Account acquisition has become more sophisticated over the years. In the beginning, card marketing consisted of relatively indiscriminate mass mailing of unsolicited credit cards to existing customers and outside mailing lists. Sensible legislation coupled with painful experience put an end to that type of irresponsible marketing. The next phase in the evolution of card marketing consisted of mailing applications to selected customers and outside mailing lists. Returned applications were then reviewed for creditworthiness and approved or declined.

As the business matured, issuers came to understand more about what constitutes a good customer and how to target that segment of the market. Industry maturity and competition brought lower response rates and an accompanying increase in the cost of acquiring a new account. From an average of between $20 and $25 in the late 1970s and early 1980s, the average cost of acquiring a new account had skyrocketed to between $75 and $80 by 1990. Acquisition costs of $100 to $130 per account are sometimes found.

Controlling the cost of acquiring new accounts is very important to the overall profitability of a bankcard program. Although most new programs achieve profitability within 24 to 36 months, when new accounts cost more than average, profitability can be delayed.

The need to continue acquiring accounts is basic to the business. In the bankcard industry, annual attrition rates of between 9 percent and 15 percent have become fairly normal. Just to maintain the current base of cardholders, a bank needs to add new accounts at the rate of 12 percent to 15 percent annually.

attrition
The loss of accounts. Attrition may be involuntary—due to bad debts, death, etc.—or voluntary—at the option of the cardholder.

Marketing Strategy

The development of a sound marketing plan is the first critical milestone in establishing a successful bankcard program. To be effective, a marketing plan must answer the four Ps:

♦ **Product.** What product will be offered and what features will be attached to it?

♦ **Price.** How much will the product cost?

♦ **Promotion.** How will the product be offered to the consumer?

♦ **Population.** Who are the best prospects for the product and how do we locate them?

Product

For ease of understanding, product is subdivided into four components: type, features, position, and design.

♦ **Type**. Will the issuer offer Visa or MasterCard or both? Will both gold and standard cards be offered? For both Visa and MasterCard? Is there a market for a corporate card?

♦ **Features**. Each issuer hopes that its card will be perceived by the market as being high value, with unique, proprietary benefits. Over the years, features that were optional, such as emergency card replacement, have become standard. There is a limited window of opportunity during which an issuer can gain market share at the expense of the competition.

emergency card replacement
A service offered by card issuers whereby a cardholder can have his or her lost or stolen card replaced within 24 to 48 hours throughout the world.

American Express was the first to offer emergency card replacement, which gave it a marketing advantage. Because it was a marketing concept with obvious merit, individual bankcard issuers, especially those with international branch or correspondent networks, copied the idea. Today it is a standard feature of Visa and MasterCard. Other examples of features that add value include extended warranty and price protection offers.

♦ **Position.** Position is concerned with the appeal of the card to various market segments. Thus, a marketer might design a standard MasterCard offering in the hope of attracting customers between 25 and 35 years of age in order to sell them additional services. That same issuer might issue a gold Visa card with features it hopes will attract affluent, high-spending, frequent travelers.

♦ **Design.** Card design has been found to be an important consideration in developing a program. The actual design is often contracted to outside agencies who specialize in this activity. The card itself is sometimes referred to as a "billboard in the wallet" and should be aesthetically appealing.

Price

The price an issuer charges for the use of its bankcard is a composite of four items.

♦ **Annual Membership Fee (AMF).** This is the annual charge imposed by an issuer on its cardholders for the privilege of carrying its bankcard. Until 1981, there were no annual fees. In 1992, the fees ranged from $0 to $25 for standard cards and up to $50 for gold cards. They typically account for about 50 percent of a program's profitability.

Market research suggests that AMF is the highest sensitivity factor for consumers. It is not unusual for the fee to be waived for an initial period of membership as an introductory offer. AT&T's initial offer of "free for life" in 1990 has had repercussions throughout the industry, although it is too early to evaluate the long-term impact of the strategy.

♦ **Interest on Outstandings.** The rate charged to cardholders is called the annual percentage rate (APR) and is calculated according to precise formulas to ensure compliance with federal and state regulations. In 1992, APRs ranged from 16.8 percent to 19.8 percent for most standard programs, with some exceptions at either end of the scale. Consumers tend to be less sensitive to APR than AMF, in part because they seldom plan to roll over their balances (pay interest).

As with AMF, it is not unusual for an issuer to offer an introductory APR that is substantially lower than the normal rate. Market research indicates that to be effective the introductory rate must be at least 3 percent lower than the rate the prospect presently pays. A cardholder paying an APR of 18 percent would probably not be induced to accept an offer from another issuer unless the APR was 15 percent or less. Consumers will readily change cards if they see an advantage that makes it worth their while. For example, assume you have a balance of $1,000 with Issuer A and are paying an 18 percent APR or $180 per year ($1,000 x 0.18 = $180). If Issuer B were to solicit you for a new account with an APR of 17 percent, would you change cards to save $10 per year ($1,000 x .17 = $170)?

♦ **Grace Period.** This refers to the time between the date of purchase and the date the issuer begins to charge interest on that purchase. The grace period is applicable only to purchases. Cash advances are subject to APR from the date of the transaction. The typical grace period is generally quoted as 20 to 25 days. In many plans, if the purchase amount is not paid within the grace period, interest is calculated retroactively from the time of purchase or posting to the cardholder account.

♦ **Additional Fees.** With the exception of cash advance fees (normally 2 percent of the advance with a minimum and maximum cap), these fees are punitive and are designed to cover additional issuer costs and to encourage the cardholder to handle the account "as agreed." The principal additional fees include

◊ late fee—imposed when the minimum payment is not received by the due date
◊ overlimit—imposed when the balance exceeds the cardholder's credit line
◊ returned check—imposed when a cardholder's payment "bounces"

Promotion

After the product has been defined and pricing has been established, the next step in the marketing process is promotion. How will the offer be made known to potential customers? How will potential cardholders be solicited? The issuer is looking for creativity and imagination in packaging the product to appeal to the target market, which includes

♦ design of solicitation material

◊ application
◊ introductory letter and brochure
◊ development of advertising copy
◊ preapproval application
◊ cash advance checks
◊ disclosure

♦ advertising

◊ selection of agency
◊ type of advertising (magazines, newspapers, radio, or television)

♦ solicitation

◊ current customers
◊ selected outside lists
◊ take-one displays in retail outlets

Population

The final P stands for population and consists of identifying the best prospects for the product. Market research plays a key role in determining which segments of the market are likely to buy a particular product.

Usage Stimulation/Retention Marketing

The marketing process does not end when a consumer accepts a solicitation. While about half of a program's profit comes from the annual fee, the rest depends on cardholder activity. The marketing function is responsible for stimulating card usage, and several techniques have proven successful. Two examples are merchandise offers mailed with the periodic statement, and the promotion of a skip payment during holiday periods.

skip payment
An offer by the issuer to permit the cardholder to omit a monthly minimum payment—generally offered no more than twice a year and only to cardholders who are in good standing. Interest continues to accrue.

With attrition rates averaging between 9 and 15 percent annually—and realizing the expense associated with acquiring a new account—it is easy to appreciate the importance of retaining accounts. Various techniques have proven effective, from courteous 24-hour customer service accessibility to periodic added-value enhancements. Extended warranty on purchases is an example of an added-value feature. It costs the issuer very little on a per-account basis, but becomes one more reason for the cardholder to continue the relationship.

Market Research

Market research helps identify those factors that tend to be most important to the target market. Good market research results in a profile of not only the target market but of the type of bankcard most likely to appeal to it. Consumers will evaluate each of the features and enhancements attached to a particular bankcard in terms of what is most important to them. Consumer A, for example, may want a card that offers a large credit line even if that means paying a higher interest rate. Consumer B, on the other hand, may view card enhancements as frills and will be attracted primarily by a low interest rate. Consumer B will probably not be motivated by loyalty to a particular bank or bankcard. In reality, of course, consumer choices are affected by more than just one factor, some not easily categorized. Following are a few of the factors consumers consider when choosing a bankcard issuer:

- ◊ price (interest rate)
- ◊ size of credit line
- ◊ length of time with bank
- ◊ other services offered by issuer
- ◊ convenience

◊ product enhancements (for example, replacement of stolen property, or lowest price guarantees)

◊ public image (for example, some consumers will choose issuers who support ecological programs)

These factors are also considered in market research so that the right product is offered to the right market at the right time and at the right price.

Summary

The marketing function is critical to the successful introduction of a bankcard program as well as to its continuing growth and profitability. As competition for customers increases, the need to monitor and revise marketing strategy becomes more important. The higher costs of attracting new accounts, coupled with the problems of attrition, make it necessary for the successful marketer to become more responsive to customers' wants and needs. Market research, both before launching the product and periodically afterwards, is a key tool in determining how the customer feels toward the card issuer and what he or she wants from the card issuer.

Review Questions

1. Name and briefly describe the four Ps of marketing.

2. Is the cost of acquiring new accounts important to the profitability of a bankcard program? Explain.

3. What is market research?

4. Are usage stimulation and retention marketing important to a program's success? Why?

5. What percentage of profit does the annual membership fee generally contribute to a bankcard's overall profitability?

5

THE CREDIT PROCESS

The effectiveness of the credit function continues to be a cornerstone of every successful bankcard program. The extension of credit must be balanced with management's prudence in determining the amount of risk taken into the portfolio. Every extension of credit carries some level of risk. There is no absolute guarantee that the cardholder will handle the relationship according to the terms of the credit agreement. Cardholders sometimes lose their jobs, spend more than they make, or borrow more than they can repay. The bank's credit department must attempt to estimate risk so that management remains informed and can plan for growth accordingly.

In this chapter, we review the credit application, its design, and methods of processing. Credit approval is discussed using a judgmental system and a credit-scoring system. We also look at performance measures and quality controls. Once accounts are on the books, they must be periodically reviewed so that credit may be continued or lines extended. Surveillance is key to preserving the overall quality of the portfolio.

After successfully completing this chapter, you will be able to

♦ discuss the objectives of the credit function

♦ explain the differences between the two major forms of consumer credit

♦ compare the judgmental and credit-scoring methods of application review

♦ explain the continuing role of the credit function after an account is approved

General Credit Policies

In all banks, the credit function for bankcards must conform to the bank's general credit policies, which govern all lending activities of the bank. These policies provide guidelines concerning the level of risk and credit criteria in the various loan portfolios. The policies also establish limits for the bank's level of contingent liabilities, which is the amount of credit available to be borrowed.

contingent liability
The total amount of credit available to borrowers but not in use. Example:

Credit line	= $1,000
Amount in use	= $ 600
Contingent liability	= $ 400

The amount constitutes a liability because the bank has committed the total amount of the credit line to the borrower. If all the bank's borrowers were to use up the total of all their credit lines, the bank would be required to extend this credit in aggregate. This includes the total amount of contingent liability plus the amount of credit currently in use.

In some banks, the general credit policies are established in conjunction with the bank's asset and liability management policies. These policies govern such areas as the funding sources for different types of loans and the maximum amounts that various classes of loans (for example, real estate, revolving credit, or fixed-rate direct consumer loans) can occupy in the bank's overall loan portfolio. The policies also govern the maximum amounts that specific deposit and other liability classes can occupy in the bank's overall liability structure. For example, a bank's asset and liability policies may establish a maximum size for the bankcard portfolio. Other policies may stipulate funding requirements for the portfolio, for example, requiring that 80 percent of the total bankcard credit outstandings be funded with long-term, fixed-rate liabilities (3-year certificates of deposit at x percent interest, for instance), with the remaining 20 percent funded with short-term liabilities (such as 30-day CDs at y percent interest).

The credit function for bankcards does not operate in a vacuum but is part of a larger bank picture and is subject to the provisions of the bank's general credit policies.

Types of Consumer Credit

The introduction of bankcards has had a profound effect on the consumer credit process. Before bankcards, most consumer loans were made on a secured, installment basis. Each time a consumer wanted to borrow money, he or she had to reapply to the bank and go through the application and approval process again. Unsecured revolving credit was limited for the most part to corporate lending and major retailers like Sears or Macy's.

A secured installment loan is one in which the bank lends a specified amount of money which is to be repaid, usually in equal monthly installments, over a specified term. The loan is secured by collateral—either the item to be purchased or another asset of equal value (such as a savings account).

Unsecured, revolving credit involves establishing a credit line of a specific amount, which can be borrowed against, in part or in full, at any time. The line is secured only by the applicant's signature and is repayable over time. The minimum payment generally runs between 3 percent and 5 percent of the outstanding balance.

Bankcard credit differs from installment lending in the following ways (as illustrated in exhibit 5.1):

♦ Because it is unsecured, the bank does not have recourse to a specific collateral if the customer defaults.

♦ The bank's exposure is always equal to the credit line, while with install- ment lending, the bank's exposure decreases each month the loan is in force.

♦ The repayment cycle, and therefore the term of the loan, is extended each time the cardholder accesses his or her credit line.

EXHIBIT 5.1   Revolving Credit Exposure

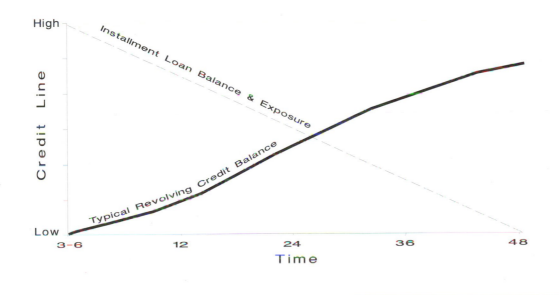

The Credit Application

The conventional bankcard relationship between a consumer and a bank begins with the credit application. The physical design of the form has to satisfy marketing, disclosure, and processing requirements. Marketing considerations include the overall attractiveness of the application and promotional material and its ease of completion. The application must not discourage the consumer because it looks complicated and cumbersome. It must call for the necessary information to be able to process it and yet be well organized and eye catching.

Many banks use a miniapplication in cases where some information is already available about the applicant. The miniapplication is easier for the consumer to complete, but it provides the bank with less detail concerning the applicant's financial profile. The miniapplication draws supplemental information from such sources as the bank's customer files, credit bureaus, and mailing lists.

Federal regulations require disclosure before the first use of the card. Regulation Z disclosures set forth the specific interest rates; the method in which the finance charges are calculated; annual fees and other charges, such as late fees for payments received after the payment due date; and the terms of the extension of credit, such as the repayment schedule and frequency of account renewal. When the consumer signs the application and the bank approves the account, a contract between the two is established. Both are committed to the bankcard relationship as long as the specific terms of extending credit are met.

The design of the application must also take into consideration the processing flow. Whether in a manual environment or an automated system, the format of the application should facilitate its ease of processing. The information requested certainly must be relevant to the bank's credit decision, but the sequence of information should also be structured to make the review and input functions easier. Let's examine the application layout of exhibit 5.2.

This application is for a bank that issues both MasterCard and Visa cards. The following is a detailed description of each section.

1. This section asks the applicant to specify the type of credit applied for: one card, two cards, or a credit line increase. There are special instructions pertaining to marital status and the requirement under Regulation B allowing married persons to choose between an individual or joint account.

2. This section calls for basic personal information: name, address, and so on. In processing, the name and related information will be the first elements entered into the automated system.

EXHIBIT 5.2 Credit Card Application

VISA/MASTERCARD APPLICATION   8

FOR BANK USE ONLY	ACCOUNT NUMBER	C L	NO. OF CARDS	CR
VISA				
MASTERCARD				
OFFICER #	INITIALS	A.B. AGENT #	**0200**	

IMPORTANT: Read these Directions before completing this Application.

Check Appropriate Box

1

☐ If this is an application for an individual account and you are relying on your own income or assets and not the income or assets of another person as the basis for repayment of the credit requested, complete only Sections A and C and sign this application.

☐ If this is an application for a joint account that you and another person will use, complete all Sections, and both parties should sign this application.

Application is for: ☐ New Account   ☐ Line Increase   ☐ VISA   ☐ MasterCard

VISA/MasterCard Experience: ☐ None   ☐ Former   ☐ Present   Account No(s) _____

I desire Credit Life Insurance   ☐ YES   ☐ NO

SECTION A—APPLICANT

FULL NAME		BIRTH DATE	SOCIAL SECURITY NO.	U.S. CITIZEN YES ☐ NO ☐	NO. OF DEPENDENTS

2

HOME ADDRESS	STREET	CITY		STATE	ZIP	YEARS THERE	HOME PHONE

OWN ☐ RENT/MORTGAGE MONTHLY RENT ☐ PAYMENTS $	IF OWNED JOINTLY ☐ SOLELY ☐	VALUE OF HOME	MORTGAGE JOINTLY ☐ SOLELY ☐	AUTO OWNED JOINTLY ☐ SOLELY ☐

FORMER ADDRESS	STREET/CITY/STATE/ZIP	YEARS THERE	BUSINESS PHONE

3

EMPLOYER	ADDRESS	OCCUPATION/RANK/RATE	YEARS THERE
FORMER EMPLOYER	ADDRESS	OCCUPATION/RANK/RATE	YEARS THERE

ANNUAL GROSS INCOME $ Other income: Income from alimony, child support, or separate maintenance payments need not be revealed if you do not choose to have it considered as a basis for repaying this obligation	SOURCE OF OTHER INCOME	OTHER INCOME $

BANK WITH	ADDRESS	CHECKING NO.	SAVINGS NO.

NAME OF NEAREST RELATIVE NOT LIVING WITH YOU	ARE YOU OBLIGATED TO MAKE MO. ALIMONY, CHILD SUPPORT OR MAINTENANCE PAYMENTS? ☐ YES ☐ NO $

ADDRESS OF RELATIVE	RELATIONSHIP	HOME PHONE

SECTION B—CO-APPLICANT

FULL NAME		BIRTH DATE	SOCIAL SECURITY NO.	NO. OF DEPENDENTS

4

HOME ADDRESS	STREET	CITY	STATE	ZIP	HOME PHONE	BUSINESS PHONE

EMPLOYER	ADDRESS	OCCUPATION	YEARS THERE

ANNUAL GROSS INCOME $ Other income: Income from alimony, child support, or separate maintenance payments need not be revealed if you do not choose to have it considered as a basis for repaying this obligation	SOURCE OF OTHER INCOME	OTHER INCOME $

FORMER EMPLOYER	ADDRESS	OCCUPATION	YEARS THERE

5

BANK WITH	ADDRESS	CHECKING NO.	SAVINGS NO.

NAME OF NEAREST RELATIVE NOT LIVING WITH YOU	ARE YOU OBLIGATED TO MAKE MO. ALIMONY, CHILD SUPPORT OR MAINTENANCE PAYMENTS? ☐ YES ☐ NO $

ADDRESS OF RELATIVE	RELATIONSHIP	HOME PHONE

SECTION C—CREDIT EXPERIENCE—APPLICANT AND CO-APPLICANT

6

Current obligations including banks, finance co., dept. stores and credit cards. Indicate debts of applicant with "A" and debts of co-applicant with "B". Indicate debts on which both are jointly obligated with "J".	A B J	ORIGINAL AMOUNT	BALANCE OR DATE PAID	MONTHLY PAYMENT
AUTO MAKE YR-MODEL WHERE FINANCED				
CREDITOR NAME ACCOUNT NUMBER				

VISA/MasterCard has an annual membership fee of $15.00.

Has either applicant ever been adjudged a bankrupt or have any judgments, repossessions, garnishments or other legal proceedings ever been filed against you? ☐ YES ☐ NO   Is either applicant liable on debts not shown? ☐ YES ☐ NO   Has either applicant ever obtained credit under any other name? ☐ YES ☐ NO   If yes, under what name(s) _____

I/We authorize the Lending Institution to make whatever credit inquiries that it deems necessary in connection with this credit application or in the course of review or collection of any credit extended in reliance on this application. I/We authorize and instruct any person, including but not limited to all local, state and federal government agencies, or consumer reporting agency to complete and furnish to the Lending Institution any information that it may have or obtain in response to such credit inquiries and agree that such information, along with this application, shall remain the Lending Institution's property whether or not credit is extended.

all information set forth in this application is declared to be a true representation of the facts, made for the purpose of obtaining the credit requested, and any willful misrepresentation on this application could result in criminal action.

7

_____   _____   _____   _____
Applicant Signature   Date   Co-Applicant Signature   Date

_____   _____
Authorized User   Relationship

3. Notice that this section calls for disclosure of the number of years in the present job. This is an important consideration for most banks in the extension of credit. Monthly salary is also included here.

4. If a joint account is being applied for, this section requests information concerning the co-applicant.

5. This section requests information about accounts the applicant may have with other banks.

6. Credit history is obviously critical. Note that the amount of current debt (the total monthly loan payments) and the total amount of current debt are requested. These factors weigh heavily in most bank lending decisions.

7. This section requests a certification that the information provided is correct. It further authorizes the bank to verify credit and employment history and evidences the applicant's agreement to abide by the terms of the credit agreement with the bank if the application is approved.

8. This section is for internal processing at the bank.

Credit Decision Process

The credit approval process is designed to extend credit to as many applicants as possible without taking excessive risk into the bankcard portfolio. Banks use different criteria for approving or declining applications. These criteria include such items as the amount of monthly income, current debt, length of residence, home ownership, and frequency of recent applications for additional credit from various lenders. Banks typically establish minimum standards for credit approval (such as minimum monthly income), but the specific standards may vary from bank to bank. Of course, Regulation B stipulates that no bank can employ standards that discriminate against applicants on the basis of race, color, sex, age, religion, national origin, marital status, or income from a protected source (any public assistance program).

Regardless of the specific standards or approval criteria, a fundamental assumption underlies every extension of credit: history will, for the most part, repeat itself. Therefore, an account that was handled properly in the past will probably be handled properly in the future. Likewise, an account that turned bad in the past is more likely to do so in the future. This means that approval criteria seek to distinguish between probable good applicants and probable bad applicants, based upon the bank's experiences with its own accounts and the credit history of the applicant. Applications, and therefore applicants, show characteristics that form general profiles. These profiles either pass or fail the bank's approval criteria.

Consider this example.

> ABC Bank has received three applications for bankcards. Applicant A, with a monthly income of $2,000, owns a home and has lived there for six years. In addition, the applicant has worked in a current job for nine years, has several credit relationships (department store and other bank accounts) with no derogatory information listed on the credit bureau report—the credit relationships have been handled properly.
>
> Applicant B, with a monthly income of $1,200, rents a home and has lived there for seven months. The applicant has taken a new job within the past year, has several credit relationships with five reported delinquencies (late payments), and made three applications for additional credit within the last year (as shown on the credit bureau report).
>
> Applicant C, with a monthly income of $5,600, rents a home and has lived there for 2½ years. The applicant has worked with the present employer for 14 years and shows a declaration of bankruptcy 3 years ago. Before this, there was no derogatory information listed on the credit bureau report. The applicant has made no recent applications for additional credit.

Though greatly oversimplified, these applicants fall into three profiles. Your intuition suggests that Applicant A's request will be approved because ABC Bank's past experience with similar applicants has been good. Applicant B's request will be declined because the bank's past experience with similar profiles has been bad. (Or the ABC Bank may require a minimum monthly income of $1,500 because experience has shown that accounts of individuals with a lower monthly income generally become delinquent.) But what about Applicant C's request? The bankruptcy of 3 years ago could result in an automatic denial because ABC Bank's experience has shown that any borrower with a history of bankruptcy cannot handle credit properly. On the other hand, the bank's experience may indicate that a 3-year-old bankruptcy alone is not a strong indicator of current creditworthiness, provided that income is relatively high and the length of time on the current job covers 12 years or more.

The point is that any bank's approval criteria are based upon its experience with different credit profiles. The bank assumes that what was generally true in the past will probably be true in the future. Different applicants' credit profiles carry different presumed levels of risk. When a profile carries too high a risk level (based upon the bank's experience), the application will be declined.

Prescreening

In the early stages of the bankcard business, the majority of applications came from "take-one" boxes located at bank branches and merchant outlets. Today, most applications are generated as the result of direct mail solicitations. Solicitation costs have increased, and response rates have declined as the industry has matured. Prescreening is a technique employed by card issuers to minimize the solicitation, and therefore the processing expense, of applicants that have little chance of being approved. The issuer runs a list of potential applicants through a credit bureau and eliminates from consideration individuals who do not meet basic presolicitation criteria. Those who respond to the promotion would then be subjected to a full credit review. The legality of prescreening has recently come under scrutiny by various agencies. The current interpretation of the law governing prescreening (the Fair Credit Reporting Act) holds that it is permissible as long as a firm offer of credit is made to everyone who passes the prescreening test (see chapter 11).

A further refinement of this type of solicitation is the preapproved credit card. The issuer sends a miniapplication—often in the guise of an acceptance form—to prospects whom the issuer has determined to be worthy of a given line of credit. Preapprovals are based on the issuer's knowledge of the prospects' credit history, stringent screening parameters, or a combination of both. On receipt of the signed acceptance, the issuer is committed to issue a card for the offered line of credit. The Truth in Lending Act (Regulation Z) covers solicitation disclosure requirements and is also covered in chapter 11.

Determining Creditworthiness

Banks use two basic methods for determining creditworthiness (or assessing the bank's risk level): the judgmental method and the credit-scoring method.

Judgmental Method

This is the oldest form of determining creditworthiness. Under this method, every application is reviewed by a bank loan officer or a credit analyst. In essence, the loan officer uses the profiles discussed earlier to arrive at a decision. On the basis of training and personal experience, the loan officer attempts to predict the likelihood that the applicants will be good or bad credit customers. We should emphasize that the credit decision is made at a point in time, whereas the credit relationship exists over time. Few applicants consciously request credit knowing they will become delinquent or cause the bank to charge off the amounts they owe. On the contrary, most believe themselves to be creditworthy. Thus, the bank depends upon the loan officer's judgment to protect it and the applicant from undue financial risk.

Demands upon the loan officer can be intense and diverse. In high-volume operations, loan officers may be required to review and pass judgments upon numerous applications each day. Like other people, loan officers have good and bad days. On any given day, they may undergo personal stress or frequent interruptions. They may need to apply new approval criteria or to interpret vague management policies calling for "tight" or "loose" credit approvals. It is virtually impossible for loan officers to precisely estimate the degree of risk taken into the bankcard portfolio through the applications they approve each day. The real consequences of their decisions will not be experienced until months or years have passed. Because loan officers are human, they can overreact if criticized for being "too loose" and decline what would otherwise be good applicants. The reverse can be true as well.

To preserve the quality of decision making in the judgmental system, management must take several steps. First, managers must remember that every credit approval carries some level of risk. Management can overreact too. The better management strategy is to establish a range (perhaps expressed as a ratio of good to bad accounts over time) within which loan officers are expected to perform. Management must also ensure that necessary information, such as credit bureau reports, is easily accessible and reliable. Adequate training and seasoning on the job are critical. A new loan officer must be given time, training, and guidance to develop the skills of effective credit judgment.

Workload is another key consideration. An overburdened loan officer cannot perform at the highest level of effectiveness day after day. Management policies must be explicit. With 20 loan officers, a tight credit policy will receive 20 interpretations. The policy should, on the other hand, address specific approval criteria, such as a higher minimum monthly income or number of new requests for credit within the past year as shown on the credit bureau report, that disqualify the applicant.

At the same time, bank management must monitor the credit judgments of loan officers on an individual basis. Typically, each loan officer is assigned a personal code that stays with the person as long as he or she is employed by the bank. This personal code is entered on every application the loan officer reviews, whether approved or declined. This allows an individual history to be regularly maintained and reviewed with the loan officer. Both management and lending personnel must be kept informed of the quality of the loan officer's credit judgment. Because the behavior of approved accounts will not be known for a period of time, management must look for signals that indicate a loan officer's performance. In the case of poor performance, remedial actions must be expeditious—whether additional training, supervision, or outright removal from a lending position. A large number of bad decisions can be made over a short period of time.

Credit-Scoring Method

The function of credit scoring is to distinguish between applicants that are expected to be good accounts and applicants that are likely to be bad accounts. As in the case of the judgmental method, scoring develops profiles that can be expected to behave in a consistent manner over time. However, the scoring method draws profiles from the bank's lending experience using statistical techniques. Credit scoring can be used in either a manual or automated environment; high-volume operations typically process and score applications with automated systems.

The language of the law governing the use of credit-scoring methods requires that they be statistically sound and empirically derived. Thus, the statistical methods used for analysis must be valid, while the values attributed to the variables in the credit scores must accurately reflect what has happened in the past. Like the judgmental method, credit scoring rests on the premise that history generally repeats itself. Although a full understanding of credit scoring is beyond the scope of this book, let's look at some of the basics.

The first step in developing a scoring system is to define good and bad accounts because precise definitions vary from bank to bank. The following are sample bank definitions for good and bad accounts:

♦ A good account is one that will not exceed a 60-day delinquency (60 days past the payment due date) and will not be delinquent more than twice during any 12-month period.

♦ A bad account is one that will become 90 days or more delinquent and may have to be charged off as a loan loss.

♦ In some analyses, a third category is defined as an indeterminate account. This is an account about which too little is known to classify it as either good or bad.

Using these definitions, databases are then identified for statistical analysis and creation of profiles, which will include financial and other personal information. In addition, they hold information from the original credit bureau report, from a sampling of good accounts currently on the books to bad accounts in advanced stages of delinquency or already charged off. Samplings may also be taken from current applications for both approved and declined applications. These samples help validate the profiles that are being created from old or charged-off accounts.

Once the samples have been selected, all the financial, personal, and credit bureau history available on each account in the sample is collected and analyzed. The purpose of the analysis is to determine the predictive value of individual elements of financial information on the accounts in the sample and the predictive value of the interrelationships among various individual elements. The predictive values are then weighted according to their elements and interrelationships. Let's see how this works.

One element of financial information is income, indicated on the original application. All accounts in the sample are analyzed at various income levels to see if a correlation exists between income level and account stability. The correlation becomes the predictive value, or weight. For instance, an income of $3,000 per month may carry a statistical weight of 10; an income of $5,000 per month may carry a weight of 15. Similarly, a bank savings account may carry a weight of 3. The idea is to isolate the correlation between the information element and account behavior over time to establish the predictive value.

Each element of financial information (for example, income, home ownership, and job tenure) is called an application characteristic. Suppose the income characteristic of $3,000 per month has a weight, or predictive value, of 10. Once the weights for individual characteristics have been determined, the relationships among various characteristics are then analyzed to determine the predictive value of specific relationships. Let's look at an example.

Characteristic	Individual Weight	Combined Weight
Income at $3,000 per month	10	
		15
Bank savings account	3	

This data shows that the applicant's income level alone has a stronger predictive value than having a savings account alone. However, customers with incomes of $3,000 per month who also have a bank savings account are more likely to handle credit in a specific manner—that is, the predictive value of the combined relationship between the two characteristics is greater than either value individually. Let's consider another relationship.

Characteristic	Individual Weight	Combined Weight
Income at $3,000 per month	10	
		11
Home ownership	12	

This particular data shows that home ownership alone has a stronger individual predictive value than when combined with income of $3,000 per month.

This analytical process examines individual characteristics and the relationships among them for each account in the sample. The various combinations are eventually sorted into profiles that make up scoring tables. These tables are the frames of reference for new applications. When a new application is processed, the system will match as closely as possible the application characteristics (individually and in relation to each other) on the new application with a profile in the scoring table. Thus, characteristics receive individual weights that add up to a credit score for the new application being processed. This score predicts the likelihood of the account becoming good or bad over time. This prediction can be expressed statistically as the estimated level of risk for each applicant. Let's use another example.

ABC Bank wants to optimize the ratio of good accounts to bad accounts. (Remember that scoring predicts the likelihood of account behavior, not the certainty of expected behavior for each new applicant.) The bank assigns a cutoff score of 200, for instance, which it determines carries odds that at least 10 good accounts will be approved for each bad account that is approved.

Applications with credit scores higher than 200 would carry more favorable odds. For example, a score of 210 may have a ratio of 15 good accounts for each bad account. Applications with credit scores less than 200 would carry less favorable odds. (A score of 190 may have a ratio of 6 good accounts for each bad account.) Therefore, to maintain a ratio of at least 10 good accounts to 1 bad account, the ABC Bank will approve all applications with a score of 200 or more and decline all applications with a score of 199 or less.

As you can see, scoring systems can more precisely predict the effects of relaxing or tightening loan policy. Some banks may want to increase the number of applications they approve. For example, ABC Bank has a cutoff score of 200 (a ratio of 10 good accounts to 1 bad account). At this cutoff score, ABC's experience is that an average of 3 out of every 10 applications are approved, giving the bank a 30 percent approval rate. Management, however, wants to expand the bank's account base by increasing the approval rate to 40 percent. A system matrix can be used to show how changes in approval criteria, for instance, will affect the portfolio. The matrix might show that an increased approval rate of 40 percent would reduce the ratio of 10 good accounts to 1 bad account to 7 good accounts to 1 bad one. Thus, the level of risk would increase. Financial analysis of the different ratios can assess the impact of the new approval rate on profitability for the bankcard portfolio, and management can thus make an informed, rather than intuitive, decision.

A few other important considerations associated with the credit-scoring method should be mentioned. First, the scoring tables should be updated regularly. The bank's experience changes over time, as do people and their financial behavior. Regular analyses for updating will preserve the integrity of the method's predictive capabilities. Scoring systems also have the capability to alert management to overall changes in the characteristics of new applications coming into the bank. If the bank's new applications are very different from the current profiles in the system, predictive capability is compromised. That is, if new applicants' profiles cannot be reasonably matched with the profiles of existing customers using the scoring tables, then the basis for predicting the account behavior of these new applicants is no longer valid. The scoring system in this instance has to be revalidated or updated.

The use of scoring can also reduce the bank's credit bureau expense—in some cases, significantly. For example, if an application score is so low that a good credit bureau report will not raise it to the cutoff level, then there is no reason for the bank to spend money for a credit report. The following sample calculation illustrates whether a bank should request a credit bureau report.

Application's score before credit bureau report	150
Maximum value of a good credit bureau report	+20
Total potential score	170
Bank's cutoff score	200

A request for a credit bureau report is unnecessary because the total potential score for this application does not meet the bank's cutoff score.

Finally, the credit-scoring method is highly objective in the eyes of the law. Professional credit-scoring firms use legally sound statistical methods for establishing their scoring tables. Because the samples used for analysis are the bank's real accounts (that have demonstrably behaved in specific ways as good or bad accounts over time), the scoring tables are empirically derived—that is, drawn from observable experience of the bank's portfolio.

One limitation of the scoring method, however, is that it cannot be applied to all types of loans. For example, a bankcard scoring system should not be used to process applications for automobile loans. To score automobile loans, the bank has to follow the analytical procedure to build scoring tables for the automobile lending function.

From management's perspective, the scoring method reduces the likelihood of discrimination in the credit function. It helps the bank approve a higher percentage of good accounts and decline a higher percentage of bad accounts. In either case, applicants receive consistent evaluation. Credit scoring also helps management assess more precisely the consequences that differing approval policies carry for the level of risk taken into the bankcard portfolio.

Automated Processing of Applications

Banks that receive many applications typically process and score them with automated systems. Credit-processing personnel use data entry terminals to enter information from each application. A preliminary score may then be rendered to determine whether a credit bureau report is needed. If so, the processing system is linked online to one or more credit bureaus. Credit records are fed directly from the bureau to the application processing computer at the bank. With the bureau information included, the system renders a final credit score. If this score is at or above the bank's cutoff score, the application is approved. If not, the application is declined. Some banks use automated application processing for handling applications but use a judgmental method for the final credit decision.

If the request for credit is approved, the automated system assigns the applicant a bankcard account number and sends a message to the bankcard masterfile to open the account. Personal information accompanies the message. If the application is declined, the system automatically prints a letter of declination to be sent to the applicant. The letter lists (in order of importance) up to four reasons for the denial of credit.

In compliance with Regulation B, a bank must respond to the applicant within 30 days from the date the application is received. If declined, the reasons must be clearly stated. Throughput rates (the number of applications processed by an operator each hour) and error rates must be monitored to ensure that productivity and quality control standards are being met. Finally, if the cutoff score is changed from time to time, management should monitor account behavior in the different cutoff score categories to ensure that credit quality is as expected.

The strategic focus of management at this point should be upon the rate of new account growth and level of risk taken into the bankcard portfolio. A cutoff score reduced to an extremely low level will raise the approval rate significantly, and many more applications will be approved. However, both the level of risk and loan losses will increase substantially. At the other extreme, a very high cutoff score will reduce the level of risk, but the approval rate will drop dramatically. A banker described this extreme: "The best way to avoid making any bad loans is to make no loans at all." The ideal is to find the optimum balance between approval rate and risk level—the "perfect" cutoff score. But no banker comes immediately to mind who claims to have found the perfect cutoff score.

Credit Review

The credit review function deals with accounts after they have been in use and focuses upon the account balance in relation to the credit line. This function involves credit extension, customer service, security, and collections. It may be handled by different departments, such as the overlimit department or credit exceptions department. Some of the credit review activities described in this section may also be a part of a bank's credit department or collections department. What is important to remember here, however, is that we are discussing the function of credit review.

A major management objective with the conventional bankcard is to provide cardholders with the maximum line of credit that is prudent from both the cardholder's and the bank's perspectives. In many banks, the credit review function establishes criteria for regular screening of existing accounts for credit line increases. These criteria are established on a computer file, or in smaller operations on a manual file, and may include such items as no delinquency and no overlimit within the past 12 months.

All accounts on the masterfile are screened against this special file. Accounts that meet the criteria qualify for a credit line increase. The amount of the increase can be a fixed percentage or a specific dollar amount for all accounts. Increases can be made by additions to current credit line ranges. For example, accounts with current credit lines ranging from $1,000 to $1,500 will be increased by 20 percent or by $200 to $300 for each account. Or accounts with current lines ranging from $1,600 to $2,500 will be increased by 10 percent or up to $250 for each account. Thus, the credit review function determines qualifying criteria and chooses the method for increasing lines of credit.

Not all cardholders, however, wait for an annual or other scheduled line increase. Some want to make a purchase that will raise their current balance above the credit limit. This group of customers can be handled in at least two ways. First, management can establish a policy to authorize transactions up to 110 percent of the current line. This means that the balance on a $1,000 credit line may increase to $1,100. Any merchant authorization request that raises the balance above $1,100, however, will be declined. Or, second, any merchant authorization request raising the balance above the set limit could be switched by telephone to a credit review officer. The officer quickly reviews the account history (on a computer terminal screen) and decides to

◊ approve the authorization request even though the balance will exceed the credit limit
◊ raise the credit limit to a higher level that will accommodate the authorization
◊ decline the authorization request

The credit review officer has the authority either to extend additional credit or to decline the request.

The credit review function also involves some aspects of collection. Activities on the account sometimes provide warning signals that it will become delinquent. For example, an authorization request raising the current balance above the credit limit may indicate the cardholder is experiencing financial difficulty. An overlimit condition is often the precursor of delinquency and charge-off. In the event of an overlimit condition, the credit review officer may decide to contact the cardholder directly, discuss the situation, and decide whether the account should be immediately assigned to the collection function for follow-up. A high level of cash advance activity may also indicate pending financial difficulty or delinquency. Again, the credit review officer may contact the cardholder to determine whether remedial action is warranted. A good credit review officer knows when to extend additional credit or commence collection activity. The action taken typically results from cardholder activity, a review of the account history, and in some cases, conversations with the cardholder.

Organizational Structure

The organization of the credit function varies from bank to bank. No single organizational structure is suitable for all banks. Exhibit 5.3 illustrates a sample organizational structure at a typical bank.

EXHIBIT 5.3   **Sample Bankcard Credit Department Organization**

The major credit functions should be represented managerially in any organization. They include

- ◊ credit approval
- ◊ application processing
- ◊ credit policy review and conformance
- ◊ credit review

Audit and compliance staff (in terms of compliance with federal regulations) normally report directly to the senior credit officer. This structure provides greater assurance of objectivity—these activities should not be unduly influenced by other management functions subject to audit and compliance scrutiny.

The senior credit officer should report directly to the person in charge of the bankcard area. This may be the bankcard department manager, division manager, or the chief executive officer of a bankcard subsidiary or a bankcard bank. The credit function is vital to the overall bankcard program. The relationship between the officer in charge of the overall bankcard program and the senior credit officer should be direct. This relationship will facilitate mutual understanding of and commitment to business practices that best serve the interests of the bank.

Summary

Every extension of credit carries risk, and a bankcard program must conform to the bank's general credit policies, which establish the level of risk and the criteria for making prudent lending decisions. Credit approval for a bankcard is usually accomplished through the use of a judgmental system or a credit-scoring system. In either case, the profiles of past borrowers are compared with those of current applicants and used as a basis for predicting the likelihood that a new borrower will have a good or bad account if the application for the bankcard is approved. Management's key responsibility is to balance growth of the bankcard portfolio with the level of possible risk taken into the portfolio. Faster growth involves greater risk. Thus, the tension between competition for greater market share and quality of the bankcard asset is ever present.

Review Questions

1. Name three ways in which bankcard credit differs from installment lending.

2. What is meant by the term contingent liability for a bankcard?

3. Although the judgmental method of extending credit is different from the credit-scoring method, what objective is common to both?

4. Create a simplified credit-scoring situation and explain the principle of predictability.

5. Set up and explain the criteria for a program for increasing credit lines for your bankcard portfolio.

6

CUSTOMER SERVICE

"Customer service is like the weather: everyone talks about it, but no one does anything about it." This is a variation on an old adage that used to be truer than bankers like to admit. However, as the competition for cardholders has increased—along with the cost of acquiring new cardholders—banks have come to realize the importance of customer service as a means of reducing attrition.

What do we mean by customer service? The term refers to any contact with a cardholder that is not related to lost or stolen card activity, credit decisions, or collection problems. Customer service is the focal point for all mail and telephone inquiries, requests, or complaints from a cardholder.

This chapter addresses the various types of customer service activities and discusses the principal reasons cardholders contact their bankcard issuer. We also learn some of the innovations that major issuers have implemented to improve service levels.

After successfully completing this chapter, you will be able to

♦ list the principal reasons cardholders contact their bankcard issuer

♦ explain what distinguishes good service from bad service

♦ identify customer service standards and discuss how they are measured and monitored

♦ discuss the potential impact of the customer service function on bankcard profitability

Customer Service by Telephone

Why do cardholders contact customer service representatives? Customers call generally because they have questions or complaints, not because they simply have nothing better to do. Each customer service representative needs to remember that there is nothing more important to the cardholder at that moment. Customers expect that any question or problem they have will receive the appropriate level of attention by the bank representative.

In most banks, certainly the larger operations, the customer service area is designated as a unique organizational unit (department or division) to demonstrate management's emphasis on and attention to the function. Think of yourself as the customer and consider the following scenario.

> You have just received this month's statement. As usual, you review the transactions printed on the statement to make sure they all belong to you. About halfway down the list, you see a transaction listed for $112.43 with the XYZ Oil Company in Miami, Florida. Unfortunately, you have never dealt with the XYZ Oil Company, nor have you been in Miami in over three years. What is your reaction? Are you calm or indifferent? Well, you'll get around to finding out what is going on when you have some time. It's Saturday anyway.

If you reacted like this, you are an unusual person. Your more likely reaction is one of fear that you will have to pay for something you did not buy, or at least be hassled by the bank about it. You may be angry that the transaction had to appear in the first place and you had to find out about it on Saturday when the bank is closed. By the time Monday arrives, your mood has turned aggressive—it's not *your* charge; it's not your fault. It must be the *bank's* fault, and you will want to make sure that the customer service person on the other end of the telephone line clearly understands your position. Or what if the charge were for $2,112.43 with a video company in New York (where you have never been) and your credit limit is $2,500? What would your attitude be when you called the bank on Monday morning?

You can see that the customer service area has to handle situations that can be stressful, argumentative, or even belligerent, depending on the nature of the problem and the customer's attitude. To meet the bank's objective of leaving the customer satisfied at the end of the conversation requires

- ◊ the kind of personality that can deal with stress
- ◊ proper training in knowing how to answer questions
- ◊ a management group that appreciates the nature of the job that customer service people do

To handle various customer situations, the customer service representative must be able to look at the issue from the customer's point of view—in a sense, roleplaying. And the customer service representative must do this repeatedly, sometimes handling over one hundred customer situations a day. Consequently, many cardholders believe the customer service representative *is* the bank.

Reasons for Customers' Calls

Depending on the size of the bankcard operation, the customer service department will take from several dozen to several thousand calls each working day. Every customer should be treated as if he or she were the most important caller of the day. Even though we recognize that every customer's call has its unique characteristics, most questions and concerns fall into certain basic categories. Let's consider some of these.

Disputed Transactions

This kind of call results when a customer reviews the statement and identifies questionable or unrecognizable entries. Think back to the situation earlier with you as the customer. Generally, these conversations are more likely to be volatile because customers are legitimately concerned about being billed for transactions that they did not make. Furthermore, an unrecognizable transaction may indicate that an unauthorized person has found a way to use the account. Therefore, the customer service representative has to be alert to the possibility that a fraudulent transaction has occurred and be prepared to report the matter to the bank's security staff.

In our situation, with the erroneous $112.43 transaction made with the XYZ Oil Company, the transaction could have been fraudulent. Most likely, however, the transaction was legitimately processed by a parent organization or by a third-party processor at the Miami location. The law requires that the name and address where the actual purchase was made be represented on the statement. But occasionally, a parent's or subsidiary's name and processing location will appear instead of the "Doing Business As" (DBA) name where the purchase actually occurred. The result, of course, is that the cardholder is confused.

Disputed transactions take many forms, ranging from the quality of service rendered to a product defect. In some cases, the failure to deliver a product will result in a disputed transaction. The customer service representative must determine from the nature of the disputed transaction what appropriate action is warranted. In most cases, the customer service representative will obtain a copy of the original document signed by the customer at the time of purchase. In the meantime, the amount in dispute is reclassified as a nonaccrual amount.

The transaction will appear on the statement but with a notation that it need not be paid and will not accumulate interest or finance charges until the issue has been resolved.

After the copy of the document has been received by the bank and reviewed by the cardholder, the customer service representative or another member of the staff must determine whether to reenter the item as a charge to the customer—who may now recognize it as his or her own transaction—or charge the item back to the merchant bank. The merchant bank must then resolve any dispute with the merchant.

Disputed transactions are sensitive in terms of customers' concerns, from the manner in which the disputed transactions are handled and monitored, or aged, by the customer service representative to the consumers' rights, which are guaranteed under the Fair Credit Billing Act. The initial telephone call will generally be the first indication of a problem, and the telephone service representative must be proficient in documenting the disputed item to ensure that proper follow-up occurs. Note that the cardholder must follow up the phone complaint in writing in order to enjoy protection under the law. A sample summary report of disputed accounts is shown in exhibit 6.1, which ages items in dispute.

EXHIBIT 6.1   Summary Aging Report of Disputed Accounts

Summary Items				Total
1. Range (Days)	1–14	15–29	30–44	—
2. Counts	1	3	5	9
3. Amounts	$412.16	$1,312.00	$2,796.14	$4,520.30

4. New Disputed Transactions Today	3	0	11
5. Disputed Transactions Resolved Today	2	0	6

Let's look at each specific area in exhibit 6.1.

1. **Range** indicates the number of days accounts are in dispute. A customer's initial call or letter disputing the transaction dates starts the aging process. The range increments are set by individual banks.

2. **Counts** indicate the total number of disputed transactions in each range category.

3. **Amounts** show the total dollar value for all disputed transactions in each range category.

4. **New Disputed Transactions Today** reflect the number of disputes entered into the system each day.

5. **Disputed Transactions Resolved Today** reflect the number of disputes settled each day.

Look again at the sample summary for a moment. Notice that the transactions in dispute status (numbers and dollars) grow larger in the older categories. The general expectation is that older items are more likely to be charged to operating losses. Management, in this example, should quickly determine why these disputed transactions are not being resolved.

Each bank should have a well-defined procedure for customer service representatives to follow before placing transactions into a disputed status. Sometimes, for example, the customer service representative can encourage the customer to look back at his or her receipts in an attempt to find a corresponding date and dollar amount. The customer in many instances will remember the purchase, and the bank avoids the effort and expense of going through the dispute process.

Depending on the bank, the responsibility for following the status of disputed items could rest with customer service or another functional area. Under MasterCard and Visa rules, an item can be charged back under specific circumstances and timeframes. If the timeframes are not met, the liability or financial exposure transfers from the merchant bank to the cardholder bank. Specific rules pertaining to the chargeback process can be found in the MasterCard and Visa operating manuals, so we will not cover them here. What must be emphasized, however, is that a system must be in place to ensure that disputed items are carefully monitored so the bank controls its exposure to financial loss.

Statement Not Received

Most customers are aware of the approximate time of the month when their statement should arrive and conscientiously watch for it. If the statement does not appear approximately on schedule, the cardholder may call to find out why. In many cases, the cardholder wants to ensure that a copy of the statement is received in time to make payment by the due date. The customer service representative must

◊ make sure that the original statement was mailed to the correct address
◊ determine how much time is left before the current payment is due

◊ make satisfactory arrangements with the cardholder to ensure that the matter is resolved expeditiously

It is important for the customer service representative to request that the cardholder contact him or her if a copy of the statement is not received, so that additional follow-up can be undertaken. Also, the cardholder's file should be noted so that anyone subsequently looking at the file will be able to determine what has happened.

Depending on individual bank policy, the specific treatment of this and other situations will vary. The customer always receives the benefit of the doubt and is assumed innocent until evidence appears to the contrary—if the customer claims the statement was not received, even if the current month's payment due date has passed, he or she is telling the truth. Most banks following this policy will reverse any late charges when the customer reports that the statement was not received. Obviously, if a customer frequently makes this claim, then there is cause for suspicion. But most cardholders will be legitimately concerned when they do not receive their statements on schedule.

Balance Information

This kind of call, as the name suggests, is an inquiry to determine the current balance on the account. The customer may be considering a purchase of some size and does not want to go over the credit limit. Or the cardholder may want to know if the last payment was posted. The call may also be made by an unauthorized person wanting to know how much credit can be used without exceeding the credit limit.

A critical part of the customer service telephone procedure is to determine whether the person calling is in fact the cardholder of account. If not, information could get into the wrong hands, and the bank could lose a lot of money along with customers' confidence. The methods for determining the identity of the caller vary. In some cases, banks include a section on the original application for private information, such as mother's maiden name, and ask the caller a related question. Or the bank may ask the caller the date and amount of the last payment. The assumption is that someone who has stolen the card will probably not know this kind of information. The specific form of identification is not as important as the general rule: every bank must have a procedure for ensuring that the caller is entitled to the information requested.

Payment Credit

Regardless of the quality of the payment processing function, mistakes will be made. When a cardholder has made a payment that is not properly credited to the account, the bank usually sends a reminder notice telling the cardholder the current payment is past due. Not surprisingly, cardholders are generally unhappy when this occurs. The customer service representative must then re-age the account so that the cardholder can continue to use the card while the problem is being remedied.

e-aging
Bringing an account from a delinquent status to a current status, or keeping a delinquent account in its present stage of delinquency so that it does not advance in the cycle.

Prompt follow-up, strong assurances to the cardholder, and documentation to the file concerning what took place are appropriate responses. Beyond being able to continue to use the card, the customer may be concerned that his or her credit rating is not adversely affected. The customer service representative must make sure that the departments researching and remedying the problem follow through.

If a payment is not credited to one cardholder's account, it may end up miscredited to another's. In these instances, the memory of the cardholder incorrectly credited is likely to be poorer than that of the cardholder whose payment went awry. The customer service representative must ensure that the reversal of the misapplied payment is made correctly and does not trigger another set of calls from unhappy cardholders.

Credit Amounts

These calls concern financial adjustments that should have appeared, but probably did not, on the current statement. Or, if they did appear, they may be incorrect. The kinds of transactions typically involved with credits on cardholders' statements are finance charge reversals, late charge and other fee reversals, as well as credits coming from merchants for such things as returned merchandise. Often, it is just a question of the time lag due to processing, and the credit will appear on the next statement. As you would expect, it is generally easier to get the information necessary to answer the question or resolve the problem if the credit originated from within the bank (a fee reversal, for instance).

If the cardholder is awaiting a merchant credit that has not yet come through interchange, the discussion can be tense. Though not at fault, the customer service representative is sometimes blamed for not knowing why the credit from the merchant has not yet arrived. There are procedures for remedying these problems, but some cardholders can be highly impatient until the credit appears on their statement.

Nonmonetary Changes

These calls are usually associated with address and telephone number changes that should be made to the cardholder's file. Unless the cardholder has experienced difficulty in getting the changes correctly made after an initial call, the conversation should be relatively low key.

Other Bank Products and Offerings

Frequently, the bankcard customer representative will be asked questions about other products the bank may offer. This is especially true during periods of special promotions. Management has to decide how much the customer service representative is expected to know about these other products. Many banks make sure their representatives are aware of special promotions each time they are conducted. They also make sure their representatives have the telephone numbers (and ideally the names) of other individuals in the bank to contact for answers to specific questions that fall outside the purview of the bankcard activity. It is important not to create a sense of "corporate runaround" in the mind of the customer. A well-established referral procedure within the bank will ensure that a customer is referred to the proper area.

The customer service representative answering the telephone *is* the bank as far as the customer is concerned. The bank's objective is to answer questions and resolve problems expeditiously. Otherwise, additional telephone calls will result, as well as letters that need responses. In some instances, individuals outside the bankcard area (such as the president or a member of the board) may hear about the problem. Let's consider some of the quality control measures that customer service representatives and management can follow to help provide the level of service the bank desires.

Performance and Quality Assurance Standards

It is important to view performance and quality standards from two perspectives: the customer's and the card issuer's. From the customer's point of view, these standards must adequately provide the value for which the customer pays. In other words, the customer judges the issuer based on service actually received and not on the bank's advertised promises of quality or excellent service.

From the bank's point of view, performance and quality standards should first answer the question: Is the customer getting value? Beyond this consideration, standards help management define their quantitative and qualitative expectations for performance. Standards, therefore, should relate to the task and be as specific as required. They must be clearly stated, understandable, and regularly monitored. A goal of establishing standards is to bring about a higher level of understanding among the staff and improved operating efficiencies.

Standards integration is critical to establishing standards that do not conflict or compete with one another. For example, a production standard in the processing of payments that is too high will probably increase error rates as operators strain to reach the level of expected production. As a result, work will have to be redone, with the risk that customers' accounts will be adversely affected before corrections can be made. Therefore, a production standard should be integrated with a standard of quality that defines the level of error rates not to be exceeded. This principle of integration applies to all standards.

Because the circumstances and nature of bankcard operations vary, we shall identify some general performance and quality assurance standards that will apply to most operations. Exhibit 6.2 illustrates a sample report that expresses various standards and can be used to monitor customer service performance. Because banks are different, their specific standards will differ.

EXHIBIT 6.2   Sample Customer Service Performance Report

Standard	Description	Current Performance	Industry Comparisons	Comments
Average speed of answer	Waiting time	X seconds	Y seconds	
Abandon rate	Percentage of calls abandoned	X%	Y%	
Other standards				

The standards for telephone customer service at banks include the following:

♦ **Average Speed of Answer**. This standard defines the average waiting time from the moment the call enters the bankcard system until a customer service representative answers it. This standard is usually stated in seconds because banks are concerned about the length of time their customers have to wait before their calls are answered. A faster response, however, will probably require more staff. Therefore, management must find the balance between an appropriate level of staff and a preferred speed of answer.

♦ **Abandon Rate**. This standard calls for a measurement of the percentage of total incoming calls that are abandoned. In some cases, a cardholder will hang up, or abandon the call, because of a normal interruption—the baby cries or the doorbell rings. But this standard is concerned with that group of customers who abandon the call because they are placed on hold too long. Over time, this condition becomes highly detrimental to a bank's customer relationships. If the condition becomes severe, customers will begin to write letters because they have grown impatient with the long wait on the telephone. Most telephone systems used in customer service today have the capability to monitor abandon rates.

♦ **Average Conversation Time**. This standard is particularly important for internal efficiency. Conversation wasted is money wasted. This standard defines the average time required to resolve a cardholder's problem or answer a question, but no more.

♦ **Telephone Proficiency**. The accuracy of information given to cardholders, telephone etiquette, and behavior under stress are included in this standard. Telephone monitoring by supervisors provides the most direct and reliable way to determine whether the standard is being met. Customer service representatives should be advised that their conversations with customers are regularly monitored by supervisors. But it is important not to make this threatening. Representatives should understand that call monitoring can be used to compliment good performance as well as identify and correct substandard performance. Criticizing a representative in the presence of others, or even in isolation, will be seen as a threat. The point is to ensure that any criticism that results from monitoring is constructive. Also, supervisors should make a point to openly compliment good performance.

Staff Scheduling

The scheduling of staff around workloads is important in any area of bankcards, but it is especially true in the telephone area of customer service. If you think about it, people who need to call the bank concerning their bankcard are probably going to make the call when it is most convenient for them—at lunch or a work break. And what about the bank whose customers reside in different time zones across the country? Obviously, it is important to have telephone service available when customers want to talk. Let's look briefly at exhibit 6.3, which illustrates a typical bank's calling activity for one day.

The left axis could represent tens, hundreds, or thousands of calls per hour, depending upon the size of the bankcard operation. Notice how the call activity varies from hour to hour, with the most sustained level of incoming calls between 11:00 a.m. and 2:00 p.m. This is not surprising because of the frequency of calls made during the lunch hour. (Bear in mind that your bank's experience may be different.) The point is that the heaviest staffing is required during these peak hours for the best level of customer service and staff resources. Now, let's look at the same bank's calling activity for one week as shown in exhibit 6.4.

EXHIBIT 6.3   Daily Call Activity Summary

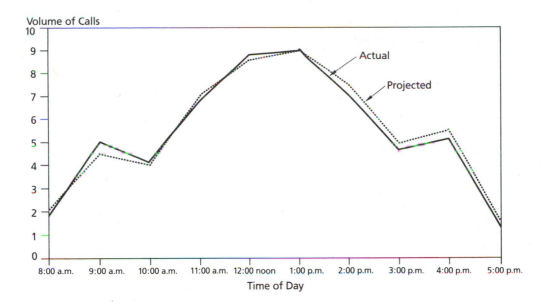

EXHIBIT 6.4   Weekly Call Activity Summary

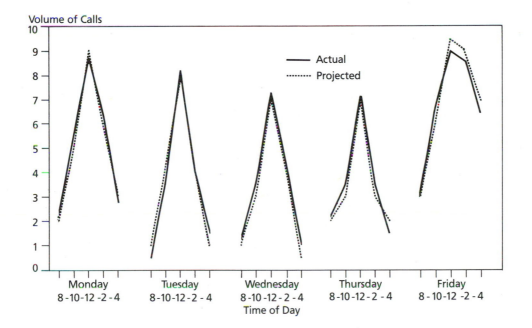

Again, we want to match resources with need. In this case, the actual and projected volumes of calls are reasonably close. This means that management can more cost effectively schedule staff. The reverse is true if actual call volumes deviate significantly from these projections.

In an effort to improve efficiency and reduce costs, many large customer service units have installed audio voice units (also called voice response units).

When a cardholder calls, the initial response is automated. A series of statements or questions are electronically generated, along with instructions on how to respond. Depending on the nature and complexity of the inquiry, the customer may be serviced with or without the intervention of a customer service representative.

Customer Service by Correspondence

Although correspondence service is obviously different from telephone service in the form of contact with the customer, the objective for both types of services is the same: Make sure the customer is satisfied. The correspondence section responds to incoming mail inquiries from cardholders. The quality of the telephone function has a direct bearing upon the workload of the correspondence group. If a telephone representative is rude, does not answer the customer's question satisfactorily, or in any manner does not properly complete the transaction, the usual response from the customer is a letter.

On the other hand, a poorly written letter from the bank, which does not resolve a matter to the customer's satisfaction, may result in a telephone call. Each function depends on the other to ensure that the customer is satisfied and that the same question or problem does not have to be addressed repeatedly.

Reasons for Customers' Letters

The reasons cardholders write are generally similar to the reasons that they call. A well-managed correspondence section involves a system for processing customer correspondence. This system includes the following features:

◊ logging in letters the day they are received
◊ classifying each letter into an appropriate category for response (an address change, for example)
◊ forwarding the letter to the appropriate person to research or take other needed action
◊ tracking each letter through the processing stream to the final response to the cardholder
◊ aging the letters in sequence to ensure that responses to customers are timely
◊ maintaining form letters to ensure correct responses to the problem raised in the customer's letter, to ensure compliance with appropriate bank policies and external regulations, and to ensure that all form letters are reader friendly (this responsibility may be assigned to the word processing area)

Performance and Quality Assurance Standards

Let's review some general performance and quality assurance standards that apply to bank correspondence programs. A report that expresses the various standards succinctly and reviews staff performance against those standards is helpful.

- **Standard Mail Response Time.** This standard assumes that an acknowledgment is sent to cardholders indicating their letter has been received and defines the average number of days for the acknowledgment. This response, depending upon a bank's volumes and policies, could be considered unnecessary. But many large operations use acknowledgments because most cardholders' letters require some amount of research and take some time to respond. This reduces the likelihood of a second letter coming in from the cardholder.

- **Time for Pending Customer Files.** The amount of time a file is pending is the amount of time the cardholder waits for the answer to a question or the resolution of a problem. This standard defines the average number of days required to resolve a customer's inquiry. The measurement begins from the date the customer's letter or inquiry is received and placed in a pending status and ends on the date the matter is resolved with the cardholder and the file is closed. The actual standard must be established within the context of a competitive level of customer service and cost of available resources—staff and related costs—to provide that level of service. Regulation Z requires that the bank acknowledge a billing dispute within 30 days and resolve it within 90 days.

- **Correspondence Accuracy.** This standard defines the accuracy of information communicated to the cardholder. This is a major concern at most banks. Erroneous information not only carries risk of litigation but also customer inconvenience and duplicated effort if the original information has to be corrected.

- **Disputed Accounts Processing.** This standard relates to the standard discussed in the telephone function and defines the average number of days required to resolve an item placed in a disputed status. This measurement begins on the date the item is disputed and ends on the date the item is resolved and notification is sent to the customer. The report for aging disputed items is key and should be carefully and systematically reviewed by management so that matters are resolved in a timely manner.

- **Suspense Clearing Time.** This standard establishes the maximum, average, or desired number of days from the date of entering a dollar amount (from a disputed or another transaction) to the day the entry is cleared. Suspense accounts must be funded by the bank and therefore must be held at minimal levels. No income is derived from a suspense account.

♦ **Payment Preparation Timeliness**. In many instances, cardholders will send their regular monthly payment in the same envelope with their letter of inquiry or complaint. Bank staffers must promptly separate and forward the payment to processing for credit to the customer's account. A standard defined more than one day in this case carries the risk that the customer will miss the payment due date. If this occurs, collection activity will begin. The customer then becomes irritated and bank resources are wasted.

♦ **Payment Preparation Accuracy**. Related to payment preparation timeliness, this standard is an example of standards integration. Payments received by the correspondence representative must be separated and forwarded with the bank's appropriate internal forms both promptly and accurately.

♦ **Address Change Timeliness**. This standard defines the number of days required to enter a change of address into the bank's computer system. If the standard is not rigorous enough, statements or renewed cards could be sent to the wrong address. Once again, the customer becomes dissatisfied and will probably call or write to the bank. This places additional demands upon the bank's telephone and correspondence staff. Furthermore, if a new card is sent to the wrong address, there is the additional risk that it could be stolen and used by an unauthorized person, resulting in more expense.

♦ **Address Change Accuracy**. Again, this standard integrates throughput quantity and quality of processed work. Sending information to the incorrect destination both inconveniences the customer and increases the bank's exposure to fraudulent activity.

While other performance and quality standards pertain, these illustrate how bankcard correspondence must be properly classified and followed through. The key is to realize that standards have an impact on the customer, the staff, and the profits and losses of the bank. Cardholders typically view their letters to the bank as the most important correspondence the bank has to handle. For most people, writing a letter signifies that the issue holds a fairly high level of personal importance. Therefore, bank customer service personnel must treat customers' correspondence with an appropriate level of attention.

Cross-Training

Cross-training is important to a smoothly running customer service operation. Correspondence representatives should be trained to assist the telephone representatives when necessary. And telephone representatives should be able to assist the correspondence representatives. The best projections will occasionally be interrupted by operating irregularities. For example, reminder notices will be inadvertently sent to customers who should not have received them. Mailing equipment may malfunction, causing billing statements to be late. These and other operating problems will occasionally cause a flood of incoming calls to the telephone section. These calls will often be followed by letters from cardholders.

When unexpected workloads hit, the ability to temporarily move staff from the correspondence to telephone section (or vice versa) will help maintain operating equilibrium and stabilize the level of service the customer receives. Some banks have contingency plans that call for training staff in departments outside the customer service area to provide backup in case of an emergency. Merchant authorization representatives could be a good alternative resource because of their existing telephone proficiency. Having a reservoir of cross-trained people qualified to handle occasional anomalies in the processing stream is fundamental to efficient productivity.

Now, let's look at how the bank handles a direct contact function with a different customer group—the merchant.

Merchant Authorizations

Merchant authorizations generally take one of two forms: terminal authorizations through POS terminals or voice authorizations. Terminal authorization is most frequently found in high-volume merchant outlets. The process begins after the cardholder presents the card. The sales person slides it through an authorization or specialized dial-up terminal that automatically calls the authorization center, keys in the sales amount, and writes down the authorization number displayed on the terminal screen. The terminal automatically dials the authorization center for each transaction. The floor limit is almost always set at zero, so that 100 percent of the transactions require authorization. This system speeds up the authorization process for individual transactions and, at the same time, helps control credit and fraud losses.

High-volume merchants often use computer central processing unit (CPU)-to-CPU linkages for authorizations. In this set-up, more processing capacity is required to accommodate a higher volume of authorization requests. An example might be an airline connected directly to major bankcard banks or third-party processors. MasterCard International and Visa International authorization systems are electronically linked because of the substantial authorization volumes they handle.

The alternative to terminal (or POS) authorization is voice authorization. Although most high-volume merchants use authorization terminals, some lower volume merchants still telephone the authorization center for credit validation. Although these merchants are not using highly sophisticated equipment to obtain authorizations, their expectations for rapid and accurate service remain high. When a merchant calls to receive authorization, the cardholder is standing there waiting, and both become quickly annoyed if the call is not responded to promptly. Thus, three factors bear most heavily upon the quality of the voice authorization function:

   ◊   adequate equipment
   ◊   trained staff
   ◊   effective work scheduling

(Many equipment options exist for technical bankcard telephone authorizations, and equipment is usually selected by staff outside the immediate bankcard authorization function. Therefore, we will not review equipment here.)

Bank authorization staff must have excellent telephone manners and protocol and be able to work efficiently with a terminal keyboard. When an authorization call comes in, the bank representative must ask for the merchant identification number (to validate the merchant in the bank's accounting system), the cardholder account number, the expiration date of the card, and the amount of the transaction for which authorization is being requested. Consider the number of computer keystroke entries for a typical voice authorization request:

   ◊   merchant identification—as many as 16 digits
   ◊   cardholder account number—as many as 16 digits
   ◊   expiration date (month/year)—4 digits
   ◊   purchase amount ($1,175.33 for example)—as many as 8 digits

As you can see, one request requires up to 44 keystrokes on the terminal before the request can be answered. Proficiency at keying will result in a prompt response.

Performance and Quality Assurance Standards

This section identifies some general performance and quality assurance standards associated with the bank's voice authorization activity. Again, the sample report for performance standards and quality assurance targets shown in exhibit 6.2 is a useful means for banks to state and monitor their standards.

♦ **Average Speed of Answer**. This standard defines the average time for the bank's authorization representative to answer the merchant's call and is measured in seconds from the time the call enters the bank's telephone system until a representative answers. Management's concern is to balance a competitive level of service with the cost of providing the service.

For example, the most competitive level of service would consist of a system large enough to handle every merchant call simultaneously with authorization representatives waiting to take all calls on the very first ring. However, at the other end of the spectrum—the most cost-effective end—the system would consist of one telephone and one authorization representative to handle as many calls as possible during an eight-hour shift. Either extreme is unacceptable; management must find the balance between competitive service levels and cost effectiveness.

♦ **Abandon Rate**. This standard defines the maximum percentage of abandoned merchant calls. The key variables influencing performance against this standard are staff levels and staff proficiency.

♦ **Average Conversation Time per Call**. Conversation time is measured from the time the call is answered by the authorization representative until the call is concluded and disconnected. This standard must be closely monitored by management. A high average conversation time simply means that more authorization representatives will be required to handle the same number of calls. This drives up the average cost per call. Industry comparisons can help in establishing the standard in this case.

♦ **Accuracy of Authorization**. Each authorization representative's activities should be logged and monitored to ensure that valid authorizations are being given. These standards should likewise include a measure of the courtesy provided to the merchant. In addition, this standard is important to monitor in terms of potential fraud. An insider at the bank, in collusion with a fraudulent merchant, could quickly expose the bank to a substantial loss by falsely approving all the transactions requested by the merchant.

♦ **Security Transfers**. Merchant authorization is one of the first lines of defense against bankcard abuse. While on the telephone, the authorization representative will occasionally transfer a merchant to the security staff because the card presented is listed as "hot." The key in this standard is accuracy. A cardholder is not likely to forgive the bank if accused of attempted misuse of the card. This situation becomes serious if law enforcement agencies are called in. Therefore, the authorization representative will be expected to operate at 100 percent accuracy under this standard.

Merchant authorization activity is clearly one of the bank's key customer contact areas. Merchants expect prompt and polite service, regardless of whether or not it is a busy time of the year, such as December, or a busy time of the day. To help balance work and resources, scheduling of the bank's authorization staff is a major consideration. Remember that the merchant authorization center operates 24 hours a day, 7 days a week. Exhibit 6.5 illustrates a sample of a bank's volume of calls requesting merchant authorizations.

EXHIBIT 6.5   Daily Volume of Calls for Merchant Authorizations

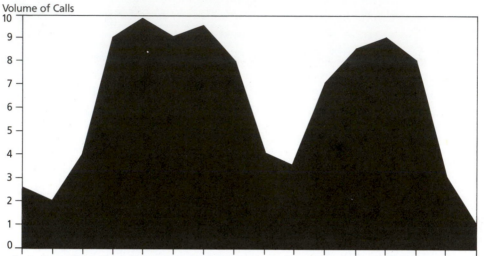

In this example, peak volumes are expected between 11:00 a.m. and 1:00 p.m. and between 5:00 p.m. and 8:00 p.m. This sample refers to one day of the week. Obviously, in a live operation, scheduling would be completed for 24 hours every day of the year. Also, some periods of the year (such as holidays, vacation season, and graduation) are typically more active than others. Staff scheduling must anticipate these unusual but busy periods.

Once the call volumes have been projected, the authorization staffing can be scheduled. By balancing staffing and call volumes, the authorization function can provide a competitive level of service on a cost-effective basis. Merchants, cardholders, and the bank all benefit from this service.

Merchant Help Desk

The merchant help desk provides assistance for merchants using dial-up authorization terminals. If the merchant has difficulty operating the terminal, the help desk staff is available to help analyze and remedy the problem. In many cases, the help desk staff can talk through a corrective procedure with the merchant to quickly repair the terminal. If a new device must be ordered, the help desk staff can place the order. Some third-party companies provide this service to merchants. So a bank must consider the costs of providing the service internally versus the costs that would be charged by a third party.

In addition to the assistance with terminals, the help desk assists merchants with many unusual situations requiring more than a routine authorization. Examples of this include extremely large dollar authorizations or confirmation of a previous authorization number. Consequently, the merchant help desk is a necessary part of most merchant bank programs.

Good Service vs. Bad Service

Good service does not necessarily mean saying yes to a customer. It does mean responding quickly, accurately, and courteously. And in cases where the issue is not resolved in the cardholder's favor, it means explaining the reasons in sufficient detail for the cardholder to understand the bank's decision. Common sense and empathy should be used when establishing operating standards. How many rings are you willing to listen to before you hang up the phone? How long are you willing to be put on hold? How many times are you willing to be transferred to have your problem solved? How often will you tolerate a sharp, discourteous, or uninterested response to your question? You have probably told friends or family about a frustrating experience with some manufacturer or service provider. Chances are you finished your story by saying, "I got so annoyed, I told them to cancel my account."

To assess customer service performance by the number or percentage of complaints about it is a common mistake. The problem with this approach is that many dissatisfied customers don't register a complaint. They simply take their business elsewhere and broadcast their dissatisfaction.

Higher Standards

Issuers have come to appreciate the value of good service and the potential damage caused by bad service. The implementation of toll-free customer service numbers and the expansion to 24-hour availability are good examples of the increasing emphasis placed on providing good service. Recent innovations like extended warranty and price protection on purchases are further examples of how major issuers have raised the level of service to their cardholders.

Summary

The three main customer contact groups in a bankcard operation are telephone and correspondence cardholder services and merchant authorizations. In the telephone and correspondence areas, it is critical to remember that the customer has initiated contact to resolve a problem or ask a question. Customers expect that the service representative will respond with an appropriate level of concern. These expectations of prompt and courteous service also hold true for the merchant authorization activity.

The bank's telephone and correspondence areas directly affect each other. A poorly handled telephone conversation will probably result in a letter, and poorly handled responses to customer correspondence will generate even more letters and incoming calls as well. Thus, establishing and applying performance and quality assurance standards will improve the level of customer service, as well as the level of customer satisfaction.

Performance and quality assurance standards define the level of service that the bank wants the customer to experience. Standards also specify management's performance expectations, both quantitatively and qualitatively. The integration of standards ensures that individual standards do not conflict with each other.

Scheduling, a major management concern, involves balancing the volume of incoming work (letters or calls) with the staff resources necessary to handle the work efficiently. Accurate projections are essential—service levels suffer with understaffing, and costs go up with overstaffing.

Management's major concerns for merchant authorization are similar to those for their customer service telephone area: scheduling to balance staff levels with work volumes so that a competitive level of service is provided cost effectively, and establishing performance and quality assurance standards that recognize customers' expectations and internal operating objectives. These challenges are intensified as the percentage of voice authorizations becomes smaller and smaller. Winning in the marketplace comes from the successful merging of customer service objectives with internal operating objectives.

Review Questions

1. Name four reasons why a cardholder might telephone the bankcard program's customer service representative.

2. How can the bank validate that the person calling customer service is the legitimate cardholder?

3. Explain the implications of performance and quality assurance standards from the perspectives of both the cardholder and the bank.

4. What does standards integration mean?

5. Describe some performance and quality assurance standards for a merchant help desk department.

7

PRODUCTION FUNCTIONS

This chapter deals with the major production functions that typically involve no direct customer contact. However, the way these functions are carried out directly affects cardholders and merchants. Point-of-sale transactions and the ongoing customer relationships can only run smoothly when the behind-the-scenes functions are performed properly.

Data processing is an integral part of bankcard operating support. Although it is a complex and technical area, our perspective will be from the layperson's point of view. We will look at distinctions between in-house and third-party processing, response time issues, capacity concerns, contingency planning, and areas of interest to management.

After successfully completing this chapter, you will be able to

♦ list the principal characteristics of production activities

♦ identify and describe the major production functions

♦ discuss the pros and cons of in-house processing vs. third-party processing

♦ cite performance and quality assurance standards for each production function

Production Characteristics

Production functions occur in large quantities and are highly repetitive. They include

◊ setting up new accounts
◊ embossing and encoding the plastic cards
◊ processing sales drafts
◊ producing cardholder and merchant statements
◊ processing cardholder payments

Because production functions involve high-speed processing, many accounts can be adversely affected within a short time if something goes wrong. Proper operating controls are fundamental to protect customer relationships, and bank error rates must be managed carefully.

Production functions have three significant characteristics. First, they involve the management of many items such as cards, sales drafts, and cardholder statements. As a result, they tend to be highly repetitive work functions. Second, if errors occur, the production functions can adversely affect thousands of customers in a single day. Third, there is no direct contact with the cardholders. Management must therefore focus on maintaining cost-effective production levels and holding error rates at an acceptable level.

Account Set-Up

Once an application has been approved, the next step in the process is the account set-up. This function involves transferring information from the application to the cardholder masterfile. Name, address, Social Security number, credit line, number of cards, and reissue cycle are among the pertinent data entered at this time. The account number is also entered, either manually or by the system. Accuracy is vitally important. This process generates a new account listing and a card embossing tape.

Card Production

The card production function transforms the blank MasterCard or Visa plastic card into a unique customer card by embossing it on the front and encoding a magnetic stripe on the back. The cardholder name, account number, and other information embossed and encoded are personal, unique characteristics of each account or customer relationship. The format, length, location, and specific information to be placed on the card are described in the operating regulations established by MasterCard and Visa. For example, the cardholder account number is assigned by the bank, but the actual length of the number is mandated by the national associations. This provides international standardization so that banks around the world can exchange data with each other. A card issued in Paris, New York, or Tokyo, for example, can be used anywhere MasterCard and Visa are accepted. Therefore, it is critical that the bank's card production function adheres to the specifications published in the operating regulations.

EXHIBIT 7.1   MasterCard® Card Design

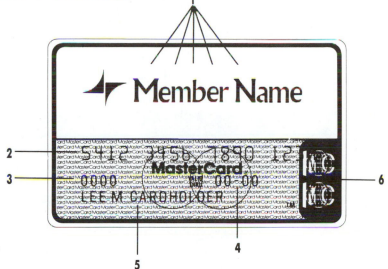

MasterCard® card reproduced with permission of MasterCard International.

Let's look at each section of exhibit 7.1, MasterCard Card Design.

1. This upper portion of the card is available to the bank for its own design but must be approved by MasterCard International. There are as many different designs as there are banks issuing MasterCard cards.

2. The cardholder account number is placed in this area. Its exact location on the card is specified in MasterCard International's operating regulations. Even the exact embossed height of the digits is specified; the last digit must always be embossed in the hologram area.

3. This number identifies the card-issuing bank.

4. The expiration date appears here. Banks can issue cards for any period of time. Some issue them for one year, some for two years, and some for six months on an initial card. The merchant bank requests this expiration date during the authorization process to ensure that a valid and current card is presented.

5. The cardholder's name appears here, on the line specified in the regulations.

6. A hologram appears here. It is designed to enhance the attractiveness of the card, to prevent tampering with the number embossed in its surface area, and to discourage counterfeiting. If someone attempts to change this number, the integrity of the hologram is destroyed. The destroyed hologram would be obvious to a salesperson, and thus the individual attempting to use the card for a fraudulent transaction would risk being caught.

Now let's look at the Visa card, as illustrated in exhibit 7.2.

EXHIBIT 7.2   Visa Card Design

VISA, the Three Bands design, and the Dove design are registered trademarks of Visa International and are reproduced with permission.

1. This area is available for individual bank design and is to the left of the Visa logo and hologram. The different card format is designed at the initiative of Visa (as is the case with the MasterCard by MasterCard International). Bank-originated designs, which appear here, must be approved by Visa.

2. The location of the cardholder account number on the card (including the embossing of digits in the hologram) is specified in the operating regulations. The same is true of the other embossed information on the card.

The magnetic stripe on the reverse of the card is shown in exhibit 7.3. The information encoded on the magnetic stripe is used in ATM and POS transactions authorized through a terminal. The card is swiped through the terminal, and the information on the stripe is transmitted through the authorization networks to validate the transaction.

EXHIBIT 7.3   Magnetic Stripe Configuration

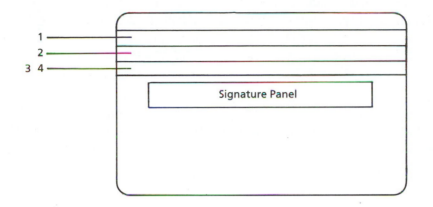

As in the case of embossing, the physical characteristics of the magnetic stripe (dimensions, location, surface finish, and profile) are specific. The magnetic stripe is composed of four "tracks," invisible to the naked eye. The tracks are International Air Transportation Association (IATA) standards on track 1, American Bankers Association (ABA) standards on track 2, and Docutel standards on track 3; track 4 is available for use by individual banks.

Card Production Process

The process that produces the finished card begins in the credit function. After an application is approved, the necessary information is transmitted to the card production function so that a card, or cards, can be embossed and encoded. The general process flow is described in exhibit 7.4, although the specific flow can vary from bank to bank.

EXHIBIT 7.4   Card Production Process

After the embossing tape has been received in the card production area, the card is produced for delivery to the customer. For the most part, the production of cards is highly automated. Sophisticated equipment embosses and encodes the plastic card in one process. Many card production facilities have automated equipment that merges the card with the card carrier form used for mailing to the customer.

Security Controls

The card production area is generally the most secure and most administratively controlled area in the bankcard operation. A fraud prevention slogan developed by the American Bankers Association in the early 1980s illustrates the need for security: A stolen bankcard is worth more than money. Consider the following example.

> John carries cash to pay for all of his purchases. The amount in his wallet is generally $500. On an out-of-town trip, John is held up and the thief takes the $500. The total value that the robber can use is the face value of the cash, or $500. When the cash is gone, the thief has to look for another victim.

> Jim carries very little cash because he uses a bankcard for most of his purchases. The same situation occurs, and the thief takes Jim's wallet with a few dollars in cash and his bankcard. Until the thief is detected through the authorization process, he can use the card up to or even beyond the established credit limit. The amount of purchases made by the thief could amount to several thousand dollars beyond this limit. So, the plastic card is indeed worth more than money within the context of card abuse.

The card production area at banks is usually isolated from other departments and activities in the bankcard operation. Access is strictly limited to authorized personnel. Every card must be accounted for daily. Any cards that are damaged in the production process must be destroyed under dual control.

dual control
A security technique that uses two or more people (or entities) operating together to protect sensitive functions, information, assets.

> Some banks institute surprise inventory checks made by bankcard officers outside the card production area to ensure that all inventory is properly accounted for and that all security procedures are followed. In some cases, banks use a separate internal mailing facility to keep the card mail segregated from other mail. This reduces the possibility that cards can be stolen before they arrive at the post office.

> Vendor security is another area of management's concern. The security function is responsible for ensuring that the plastics vendor has adequate internal controls. Once the bank's name is placed on the card by the card manufacturer, the risk of fraudulent abuse is much greater. Blank plastic cards, can easily be embossed and encoded for fraudulent use. The issuing bank is responsible for all such losses.

Card Production Controls

A convenient way to address the need for and nature of card production controls is through performance and quality assurance standards. Again, the standards discussed here are general; each bank's management will establish its own specific guidelines.

♦ **Daily Embossing Timeliness.** This standard defines the average number of days from receipt of the embossing and encoding tape to the date the cards are mailed to customers. The specific number of days must take into consideration the requirement of the Equal Credit Opportunity Act, Regulation B, concerning the time permitted for response to the applicant. In other words, after the application is received, approval (or declination) must be determined and the customer notified within 30 days. In many banks, the notification of approval occurs with the mailing of the bankcards to the customer.

A second important consideration is that of customer service. Once the application has been submitted, the customer is anxious to receive the card. A quick turnaround time from the date of application to the date of card mailing can help build customers' goodwill and support efforts to differentiate the bank in the marketplace. Turnaround time is a characteristic of the quality of customer service.

♦ **Daily Embossing Accuracy.** This standard defines the reject rate expressed as a percentage of cards embossed and encoded each day. In most bankcard operations, the embossing and encoding equipment self-checks the information on each card to ensure its accuracy. Not only is production wasted if the discard level is too high, an undetected "reject" mailed to a customer can result in embarrassment to both the customer and the bank.

♦ **Card Carrier Accuracy.** The card, or cards, are usually mailed to the customer in a card carrier form. This form is printed with the customer's name, account number, and in some instances, the amount of the credit line. Where the card and card carrier are matched manually, special care must be given to ensure that the appropriate card is inserted in the appropriate card carrier. Equipment is available to automate this activity and has a self-checking mechanism to ensure accuracy. Whether manual or automated, the standard should call for an extremely high percentage of accuracy—100 percent accuracy would not be inappropriate.

♦ **PIN Release Timing.** The personal identification number (PIN) is the unique number or code assigned to the customer for access to ATMs. For security reasons, the plastic card and the PIN are not mailed in the same envelope. If an unauthorized person were to get the card and the PIN, the card could be easily used to obtain fraudulent cash advances. Most banks mail the PIN to the customer a few days after the card has been mailed. This standard defines the average number of days between the dates the card and PIN are mailed to the customer.

♦ **Card Issues/Reissues.** Some banks use a "good from/good thru" procedure on card issue/reissue. The card is good from a specific date (effective soon after the card is mailed to the customer) through a specific date (usually one or two years following the "good from" date). This procedure helps protect the bank because a new card cannot be used before the "good from" date and is referred to as dual dating.

Standards for the accurate handling of reissued cards parallel those established for new cards. Because of customer service and security concerns, the standards must be stringent.

♦ **Cost per Card Issued.** The speed and accuracy of the card production function are key factors when considering a standard for cost per card. High reject rates, excessive equipment downtime, and idle time (the time equipment remains unused when cards could be processed) all drive up production costs. The national average is about 1.7 cards per account, so the cost per card has an impact on the overall profitability of the bankcard program.

♦ **Exception Management.** Exceptions must be carefully handled in the production of cards. For example, the credit department may decide at the last minute not to reissue a card because the relationship has deteriorated. Card production staff must pull the card from the production stream before it is mailed to the customer. On the other hand, situations arise in which a customer needs the card sooner than the routine processing time normally permits. For example, the customer is leaving for vacation and needs the card sooner than scheduled. The card production staff must see that the card is produced quickly and forwarded to the appropriate bank officer or mailed to the customer. The standards for exception management cannot accommodate errors.

Card Production Scheduling

The scheduling of card production volumes is important to spread the work over time. For example, if your bank conducts a major card acquisition campaign during the months of March and April, the consequences will be heavy card production workloads during April through June. Expiration dates on the new accounts booked as a result of the promotion should be distributed evenly over 12 months to avoid having a recurrence of the peak workload during April through June of the following year. Because some months of the year are typically more active for acquiring new accounts, failure to establish staggered expiration dates for new accounts could produce a work schedule for card production that looks like Schedule A illustrated in exhibit 7.5. Evenly distributing the expiration dates will result in reissue activity in the more desirable Schedule B, also shown in exhibit 7.5.

EXHIBIT 7.5   Sample Card Production Schedules

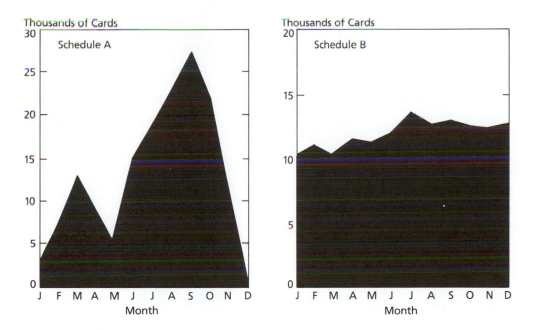

The peaks of production in Schedule A will place unreasonable demands upon the staff and the equipment. Crises are more likely to occur. For example, if a machine breaks down during a peak period, there is no time for repairs because the backlog of cards to be processed will grow at an overwhelming rate. On the other hand, during low-volume production—April/May and December—staff and equipment are idle. Production Schedule B, on the contrary, evens the workload. Staff and equipment are employed more efficiently.

The responsibility for establishing expiration dates for new accounts usually is a part of the account set-up function. Systematic distribution throughout the year will best serve the interests of customer service and internal operations.

As you can see, a great deal of effort is involved in the production of bankcards. Customer service standards, security and cost controls, adherence to specifications in the operating regulations of the national associations, and other requirements must be met. A proficient card production function generally goes unheralded, while an error-prone operation can do lasting damage to a bank's reputation for customer service. Accuracy and efficiency must be the predominant characteristics of card production.

Sales Draft Processing

The sales draft processing function converts information from the merchant draft (paper) to computer tape (electronic) so that items can be entered into interchange. When the merchant bank "sells" the sales drafts to cardholder banks through the interchange system, the specific function is called settlement. The draft processing function is necessary because paper transactions cannot be entered into interchange; transactions must be in machine-readable format.

The key elements of information on a paper sales draft that must be captured and transferred to a computer tape or disk are shown in exhibit 7.6.

EXHIBIT 7.6   Key Elements of a Merchant Draft

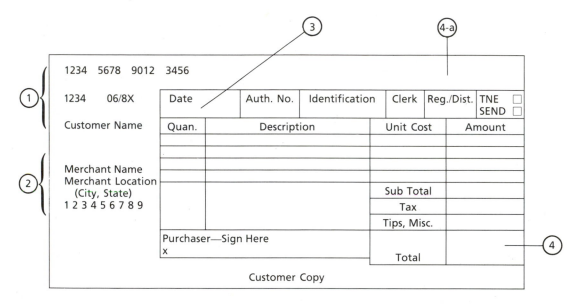

1. This is the area for the cardholder's account number and name, as well as the expiration date (used in the authorization procedure to ensure that the card is current). Only the account number is actually captured.

2. This section on the draft is for the merchant's name and identification number. Remember that the merchant's name must be identifiable to the cardholder—a different parent company name is unacceptable. In virtually all cases, the merchant information from this section already exists on the merchant bank masterfile, so that all the data elements represented here do not have to be captured during processing.

3. This represents the date of the transaction, which must be captured, passed on through interchange, and then finally printed on the cardholder statement.

4. The total amount of the transaction must be captured for settlement and billing to the cardholder. Section 4-a shows the location for imprinting the amount of the transaction on the draft.

Although the specific sequence and the exact tasks performed differ from bank to bank, the processing stream follows the steps illustrated in exhibit 7.7.

EXHIBIT 7.7   Steps in the Sales Draft Processing Stream

- ◆ **Draft Preparation.** This stage involves preparing the drafts for the capture function—by ensuring all the drafts in a processing batch are properly matched according to depositing merchant, the maximum number to be processed in a batch is not exceeded, the drafts are legible, and the like.

- ◆ **Data Capture.** This stage is where information is converted from the paper draft to tape or disk. In a manual system, the required elements of information are keyed from the individual draft into a computer terminal. An automated capture function uses high-speed equipment to capture as much information as possible as the draft passes by the sensory head on the machine. The amount of the transaction has to be in optical character recognition (OCR) format for the equipment to read the amount. The amount is imprinted in section 4a of the draft as shown in exhibit 7.6.

- ◆ **Data Completion.** This activity ensures that all the necessary information has been captured. Used primarily in an automated capture function, information that could not be read by the machine is entered through a terminal so that a fully captured transaction can be entered into interchange for settlement.

- ◆ **Balancing.** Because mistakes are made when the amounts are captured from the drafts, this activity resolves and adjusts out-of-balance batches of drafts. All batches must be in balance before their transactions can be passed on for settlement and entered into interchange.

- ◆ **Settlement.** At this point, the information from the merchant draft has been put into electronic form, all necessary information is present, and the batches are in balance. The settlement function now sells the value of the day's work to interchange and obtains reimbursement for the drafts' value from interchange.

Processing Controls

Draft processing is a cost-sensitive area to the bank. Therefore, performance and quality assurance standards must focus on the primary areas of management concern. Let's consider some standards found in most operations.

♦ **Draft Processing Timeliness.** The time within which drafts are processed has financial consequences to the merchant bankcard operation's bottom line. Most banks give the merchant credit on the same day the merchant makes the deposit. Yet, the bank cannot get reimbursement from interchange until the drafts have been processed. The bank is then in a negative float condition until reimbursed from interchange. If it takes two, three, or five days to complete the draft processing activity as shown in exhibit 7.8, the float expense to the bank rises accordingly.

EXHIBIT 7.8   Elapsed Time in Sales Draft Processing

	Day 1	Day 2	Day 3	Day 4	Day 5
Bank A	Merchant Deposit Credit				Settlement (Reimbursement)
Bank B	Merchant Deposit Credit		Settlement (Reimbursement)		

In the first case, Bank A must carry the amount of the merchant deposit in float for four days—the bank pays for its money on a daily basis too. Bank B saves two days of float expense because the draft processing function is more efficient.

The timeliness standard establishes management's expectations for draft processing time. For example, the standard may be stated as 90 percent of the drafts received at the center by 2:00 a.m. will be processed by 2:00 p.m. the same day. The lower the percentage required by the standard, the greater the float expense is to the bank.

♦ **Draft Processing Accuracy.** Speed of processing is of little value if error rates are high. This standard integrates the speed of production with the quality of the work processed. The standard should seek the maximum speed of drafts processed and the minimum level of errors in processing.

You can see that the draft processing function is a critical production activity in the bankcard operation. It is typically high volume and requires tight operating controls and performance standards. Processing time means money saved or lost to the bank.

Throughput (the amount of work processed) measures should be established for each operator to ensure that performance standards are met. These measures could be expressed in terms of average number of keystrokes expected per hour or number of drafts completed per hour. These measures should also be applied to work sections and shifts. In a production function like draft processing, the importance of speed and accuracy can hardly be overstated. Individual and group throughput measures give focus to management's expectations of speed and accuracy.

Statement Production

The printing and mailing of cardholder statements is a sensitive production activity. Not only must these statements be accurately rendered, they must be mailed to cardholders in a timely manner. Under the Fair Credit Billing Act, cardholder statements must be mailed no later than 14 days before the payment due date. Proper controls must be in place to make sure the statement production function works properly. Although the specific format of the cardholder statement varies from bank to bank, exhibit 7.9 highlights the main features.

1. This area is for the account number and allows the cardholder to quickly verify the statement. In many instances, cardholders maintain several bankcard accounts, and monthly reconciliation of receipts with these statements must be accurate.

2. This area shows the credit limit.

3. The available credit shown each month is important to ensure that the cardholder does not exceed the limit. The amount shown on the statement excludes transactions that had not reached the cardholder bank at the time the statement was produced and transactions that have occurred since production of the statement.

4. This box shows the cardholder how many days are included in the billing cycle. This number may change slightly, depending on the number of days in the billing month.

5. This is the cutoff date for entering transactions on the statement for the billing period. Transactions received by the bank after the closing date will appear on the following month's statement.

6. This is the final date by which payment must be received for the statement's billing cycle. Payments received after this date are delinquent and may be subject to a late charge.

EXHIBIT 7.9   Sample Cardholder Statement

ABC Bank

Account Number	Credit Limit ②	Available Credit ③	Days in Billing Cycle ④	Billing Cycle Closing Date ⑤	Payment Due Date ⑥	Minimum Payment Due ⑦
① 1234 5678 9012 3456	2500	2456	30	06/20/XX	07/09/XX	20.00

Trans. Date	Post. Date	Reference Number	Account Activity Since Last Statement	Amount
⑧ 0612	0615	545377ADG01RLC	Acme Restaurant, Miami, Florida	43.17
⑨	⑩		⑪	⑫

Previous Bal. −	Payments −	Credits +	Charges and Advances +	Late Charges +	Debit Adjustments +	**Finance Charge** =	New Balance
0.00	0.00	0.00	43.17	0.00	0.00	0.00	43.17
⑬	⑭	⑮	⑯	⑰	⑱	⑲	⑳

Average Daily Balance × Monthly Periodic Rate = Periodic Finance Charge

㉑				
Nominal Percentage Rate	xx.xx	0.00	x.xx	0.00
Monthly Periodic Rate	x.xx	Periodic Finance Charge + Transaction Finance Charge = Total Finance Charge		
Annual Percentage Rate	x.xx	0.00	0.00	0.00

Notice: See reverse side for explanation of finance charges and other important information.

7. The amount of the minimum payment is determined by management and varies from bank to bank. Since the mid-1980s, the trend has been to slightly lower the minimum payment due. The amount of the minimum payment, however, must be at least adequate to cover interest charges and reduce the principal.

8. This date should match the date on the cardholder's receipt. Its purpose is to enable the cardholder to verify the date and accuracy of the transaction.

9. This date indicates when the individual transaction was actually posted to the cardholder's account. The posting date is used in calculating the average daily balance.

10. The reference number uniquely identifies the transaction to the cardholder. If there is a question or subsequent dispute of the transaction, the reference number allows the bank to trace the transaction back through interchange to the merchant bank.

11. This area lists each transaction received during the billing cycle with the merchant name and location.

12. The amount of each transaction appears in this column.

13. The previous balance that was carried forward from the past month's billing cycle is shown here. If the account was paid in full last month, the amount will be zero.

14. The total amount of last month's payment is shown here.

15. The sum of any credits, such as a credit for merchandise returned to a merchant, is shown in this box.

16. The total amount of charges and cash advances received during the billing period are shown here.

17. If last month's payment was not received by the payment due date, the amount of late charges is shown. Individual banks determine the specific amount of the late charge to be assessed.

18. Debit adjustments include items such as membership fee.

19. Note that the words "finance charge" are in bold type. This is required by the Truth in Lending Act's Regulation Z. The specific amount of the finance charge for the billing period is printed in this box.

20. The number shown here is the result of the calculations across this line on the statement:

    previous balance − credits + debits + finance charge = new balance

21. This section is required by Regulation Z and shows how the monthly finance charge is calculated. Note that the nominal percentage rate is the disclosed interest rate expressed as a rate for a 12-month period. The monthly periodic rate is the nominal percentage rate expressed as a monthly rate. The annual percentage rate (APR) is the actual effective rate, after applying the monthly periodic rate plus any transaction charges.

The reverse side of the statement is important too. You will always find an explanation of the method used to calculate finance charges and an explanation of cardholder billing rights—the steps the cardholder should take in the case of questions about individual transactions or errors on the statement.

How Finance Charges Are Calculated

Banks calculate the total finance charges shown on the statement depending upon the periodic and transaction finance charges.

♦ **Periodic Finance Charges.** These are calculated in three steps and are explained on the reverse of the statement as follows:

◊ We determine the outstanding balance in your account for each day.
◊ We then add up the outstanding balances and divide by the number of days in the billing cycle. The result is called the average daily balance.
◊ We then multiply the average daily balance by x percent (the monthly periodic rate).

♦ **Transaction Finance Charges.** These charges are also explained on the reverse of the statement: We charge your account x percent of the amount paid on each cash advance posted to your account during the billing cycle. The minimum charge for each cash advance is x dollars.

Billing Rights Summary

Banks also provide a billing rights summary in case of errors or questions about the bill. If the cardholder thinks the bill is wrong or needs more information about a transaction listed, the back of the statement provides the following instructions.

Write to us on a separate sheet of paper and in your letter give us the following information:

• Your name and account number.
• The dollar amount of the suspected error.
• A description of the error. Explain, if you can, why you believe the error exists. If you need more information, describe the specific transaction you are unsure about.

Special Rule for Credit Card Purchases

The reverse of the statement also includes information such as the following:

If you have a problem with the quality of goods or services that you purchased with a credit card, and you have tried in good faith to correct the problem with the merchant, you may not have to pay the remaining amount due on the goods or services. You have this protection only when . . . .

A number for a lost or stolen card and emergency services is included on the reverse of the statement.

Statement Production Controls

Performance and quality assurance standards for the main activities are discussed below. Although we focus on cardholder statements, merchant statements must not be forgotten. Standards must be established for the production accuracy and timeliness of merchant statements that are as specific and explicit as those for cardholder statements.

♦ **Statement Timeliness.** This standard refers to the average number of days between cycle closing date and mailing date. As noted, the Fair Credit Billing Act requires that the statement be mailed no later than 14 days before the payment due date. But the bank attempts to get the statement in the cardholder's hands as soon as possible after the cycle closing date. The cardholder then has more time to review the statement and ensure that payment reaches the bank by the due date.

♦ **Statement Mailing Accuracy.** The concern here is accuracy of statement inserts (promotional brochures or other material). Bear in mind that some inserts, such as an offer to skip a payment, go to selected customers only, not to all customers. The accuracy standard is important in avoiding customer confusion and irritation.

♦ **Damaged Statements.** The tolerance level for damaged statements is expressed as a percentage of the total statements produced. Beyond the waste that results from damaged statements, each unusable statement must be redone, in some cases manually, which affects the cost of production.

♦ **Statement Repair.** This standard states management's expectations for the length of time it takes to correct a damaged statement. The standard is usually expressed as the average number of days from the date the damaged statement is ordered until the date the corrected statement is mailed. The Fair Credit Billing Act specifies deadlines that apply to this standard (a minimum of 14 days before payment due date). The requirement for good customer service also dictates deadlines. Cardholders usually know approximately when to expect their statements. When the statement does not arrive as expected, they may make a telephone call to the customer service department.

♦ **Mailing.** Statements delivered to the mail department must be carefully monitored as well. Proper edits and management controls are essential to ensure that all of the statements that should be produced are produced and that all statements that should be mailed are mailed. Exhibit 7.10 shows what the points of control (simplified) might look like, depending upon the processing stream of the individual bankcard operation.

EXHIBIT 7.10   Points of Control in the Processing Stream

As with card production, the workload should be as balanced as possible throughout the month. The number of accounts per billing cycle should be evenly distributed. If one billing cycle has 90 percent of the accounts, and therefore 90 percent of the statements to be produced and mailed, chances are that the statements will not go out on time. On the other hand, if there are 10 billing cycles and each cycle contains approximately 10 percent of the accounts, the demands upon staff and equipment will be more reasonable. Statements will be mailed each working day, as opposed to having overwhelming numbers of statements on some days and virtually none on others.

Payment Processing

The payment processing function is another area of sensitivity to cardholders and management. This area is responsible for receiving, preparing, capturing, and crediting cardholder remittances. Let's review briefly the basic steps in the processing stream.

Receipt

At first glance, little needs to be said about receiving remittances. The specific date each payment is received must be carefully noted so that the cardholder promptly receives payment credit. Payment is credited the day the bank receives the payment, even if it is actually processed at a later date. Also, cardholders frequently use the remittance coupon as a means to communicate with the bank—notes range from telling the bank about a change of address to disputing an item on the previous statement. Responses to bank promotions may be received. These must be forwarded to the appropriate departments.

Preparation

After payments have been received and logged, they must be prepared for processing. Whether the bank enters the payments manually into a terminal or uses automated reading equipment, payments are usually grouped in batches of a specific number to facilitate processing and balancing.

Capture

The next step is to capture the information from the remittance coupon and the check so that the cardholder's account can be credited for the amount of the payment. In a manual operation, the account number and amount of payment are keyed into a terminal. In a highly automated operation, the account number is read automatically, so that the operator keys only the amount of payment. Automated equipment can also determine whether the payment is below the minimum required or more than the total amount due, at which point the machine alerts the operator for special handling. The operator need only concentrate upon the speed and accuracy of keying in the payment.

Exhibit 7.11 illustrates a sample bank remittance. Although it probably differs from the forms your bank uses, it shows the kind of information captured. The highlighted areas are those captured in the processing function, which ensures the proper amount is credited to the appropriate cardholder account.

EXHIBIT 7.11 Sample Bank Remittance

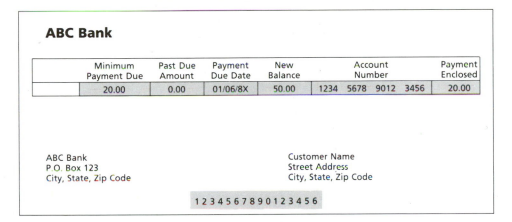

	Minimum Payment Due	Past Due Amount	Payment Due Date	New Balance	Account Number	Payment Enclosed
	20.00	0.00	01/06/8X	50.00	1234 5678 9012 3456	20.00

ABC Bank
P.O. Box 123
City, State, Zip Code

Customer Name
Street Address
City, State, Zip Code

1 2 3 4 5 6 7 8 9 0 1 2 3 4 5 6

MICR Read

Balancing

After the remittance information has been captured, the payments must be balanced to ensure that the total amounts received have actually been processed and are ready for crediting to cardholders' accounts. The balancing is usually performed for each batch of payments processed and then for the total of the batches processed for the shift or day. Errors must be discovered quickly so that adjustments can be made promptly.

Posting

Once balanced, payments are then posted, or credited, to individual card-holders' accounts. Posting is performed daily. For payments received one day or more before the posting run (such as payments coming in on a Saturday but not processed until Monday), the payments are back dated so that the cardholder's account is credited for the specific date the payment was received at the bank, as required by federal law.

Payments made at a branch bank are forwarded to the card center for processing (unless the branch has the capability to process bankcard payments). Assuming it takes at least one day for the payments to reach the card center, all of the payments will be back dated. Prompt handling of branch payments is important because some customers make the payment on the payment due date. If the payment is not processed and the cardholder's account credited quickly, a statement will be produced showing that the payment has not been received and the account is delinquent. Many card operations have a separate section and procedure for handling branch payments.

Remittance Processing Controls

Let's look at some performance and quality assurance standards for the payment processing function:

♦ **Remittance Processing Timeliness.** Payments must be processed in a timely manner. Nothing is more irritating to a customer than not to have a payment reflected on the monthly statement and to receive unwarranted late notices and late charges. This standard establishes management's expectations concerning specific processing times. One way to state this standard might be x percent of remittances or payments received by 6:00 a.m. will be transmitted for posting by 4:00 p.m. the same day.

♦ **Input Errors.** Despite diligent effort, there will be errors. But the level of errors must be strictly managed. An undetected error will annoy the customer and create additional work for the remittance function and other areas of the bankcard operation. The standard sets the expectation for the percentage of total payments to be processed correctly on the first processing run.

♦ **Branch-Originated Remittances.** This standard establishes the timeframes for handling payments made in the branches. For example, the standard may be that all payments received in the card center by 6:00 a.m. will be processed the same day. This standard should take into account that some customers will make payment at the branch exactly on the date the payment is due. If they are not processed promptly, customer dissatisfaction and additional work for branch personnel and the customer service department will result. Also, a separate standard should be established for accuracy, or quality, of the work processed.

♦ **Balancing of Remittances.** The balancing function ensures that the total amount of dollars processed through the remittance function is accounted for and in balance. The standard should call for an extremely high percentage of transmissions for posting to be balanced accurately.

♦ **Suspense Processing Timeliness.** Part of the reality of the processing stream is that difficulties will arise that cannot be immediately remedied. Items of this nature are placed into a suspense account until the problem can be resolved. Management's concern is that items in suspense stay there no longer than absolutely necessary. This standard defines the average number of days from the date the item is placed into suspense until the date the entry is resolved and cleared. Bear in mind that as items get older, they become harder to resolve.

♦ **Suspense Processing Accuracy.** Obviously, the timely removal of items from suspense is key, but so is the accuracy of clearing and balancing suspense accounts. The standard could be expressed as a percentage of items handled, a percentage of dollars handled, or both; that is, x percent of items and y percent of dollars processed through suspense will be accurately handled.

Processing Systems

Although not strictly a production function issue, various processing systems support bankcard production. Processing systems include data processing and peripheral activities. Depending on the bank, the processing systems may reside in-house, or they may be purchased from third-party processors. Many banks use a combination of in-house systems and third-party processors.

Certain issues are of concern to management regardless of whether the activity is performed in-house or by a third party. Proper management of these activities is fundamental to an effective bankcard operation. Because our focus is nontechnical, we will simplify these discussions.

In-House and Third-Party Alternatives

The processing options available to every bank are

◊ an in-house operation in which all functions are performed within the bank

◊   a third party that performs virtually all functions except those
     associated with policy formulation and direct customer contact
     activities
◊   a combination of the two

There is no prescribed formula that works best. Which processing alternative
to choose must be based on a business analysis combined with a strategic
direction. One view is that an in-house operation gives the bank more control
of its destiny, while a third-party arrangement offers less. However, an
in-house operation requires a high level of activity to be cost effective. Each
bank's particular situation must be analyzed to determine the most appropriate
configuration. For example, the bank could perform the card production
function but have the masterfile maintained by a third-party processor. Or the
bank could print statements from a tape received from the third-party processor.
The possible combinations are virtually unlimited.

There is a growing trend toward the use of third-party processors and every
indication that the trend will continue. The autonomy, competence, and
integrity of large processors have made them an attractive alternative to in-
house processing. Additionally, rapid changes in technology, and the costs of
upgrading software and hardware, give the third parties a competitive edge in
achieving and maintaining economies of scale.

Capacity Concerns

Management's primary concern is whether adequate capacity exists to handle
the daily and projected volume to some reasonable point in the future.
Depending on the bank, this could be one year, three years, five years, or some
other timeframe. The capacity of the computer hardware and software to
process the volumes must be properly assessed. If too great a capacity exists,
money is wasted because the system is idle and uses only a small portion of its
potential. If too little capacity exists, the work cannot be processed in a timely
manner. Exposure to loss increases and customer service levels decrease in
this situation. Management's projections of future volume should determine
the capacity required.

Processing Windows

Processing window describes the time in which each day's work must be
processed. For example, each day the bank must settle with interchange—sell
merchant transactions and buy cardholder transactions. If the processing
window is missed, the bank must bear the cost of float for an extra day.
Remember that most banks give merchants credit on the same day they deposit
the drafts (or send transactions on computer tape for deposit) with the bank.

Consider the consequences of missing daily file updates. The customer service department will have to use stale information because last night's payments will not be reflected. The authorization department will also be using this stale information and possibly decline a request for authorization because the removal of an account from a delinquent status will not be reflected. The collections department will not be able to determine whether delinquent cardholder "promises to pay" have been kept because yesterday's payments will not appear. Meeting processing windows regularly is critically important.

Response Times

Adequate response times must be maintained. For example, let's assume that the average response time for a simple balance inquiry in the customer service department is one second. This means that after the customer service representative has entered the cardholder account number and the command for the balance inquiry to appear on the terminal, no more than one second elapses before the information appears on the terminal.

What if this response time unexpectedly goes to three seconds? The productivity of the customer service representative is reduced to one-third of the expected response level and is beyond the control of the customer service representative. What if the response time goes to five seconds? Productivity and customer service levels are reduced to one-fifth of the one-second response level. The consequences are the same for the collection function, the credit review function, and every other function that depends on response times to maintain productivity and service levels. Although the reasons for higher and lower response times are more technical than we will discuss here, the point is that response times have a direct relationship to levels of customer service and to productivity in the bankcard operation.

System Availability

The percentage of time that the bank's computer system is up and available is another critical factor. For instance, if the system goes down during the peak hours of the day for customer service, cardholders' questions cannot be answered or complaints remedied. Customers have to call back when the system is up. Again, more work results because of duplicated calls that would have been resolved if the system had been working. Authorizations cannot be handled because the terminals are blank—operators have no access to cardholder files. Merchant authorization terminals also cannot obtain authorizations because the system is down. System availability is critical. The system must be up and available the vast majority of the time, especially during peak hours of operation.

Audit Controls

The need for proper monetary and nonmonetary audit controls in the processing system cannot be overstated. A great deal of money passes through the merchant deposit, draft capture, and remittance processing systems, all of which flow to the general ledger. The credit staff books accounts to the masterfile, the customer service staff makes address changes and reverses finance charges, the collection staff re-ages accounts, and so on. While subsystems interact with the masterfile, audit controls to validate the accuracy and integrity of the main processor are fundamental.

Contingency Planning

Federal regulations require the existence of a formal contingency plan for operations in the event of a major disaster. Your bank is required to have a contingency plan, and chances are the bankcard operation is required to have one as well. At least the data processing systems must be covered in a contingency plan.

Every major component of the operation should have an alternative means of carrying on should a major disaster occur. Duplicate paper or electronic files should provide backup for customer files maintained on the masterfile and subsystems, so that documentation to recreate payment entries, address and other nonmonetary changes, new account bookings, and the like is available. Some banks have formal arrangements with another bank's data processing function outside their immediate geographic area for a backup in the event of a disaster. A few banks are large enough to have a duplicate system operating in another geographic region that would serve as a backup in an emergency situation. Each department of your bankcard operation has to be able to function in an emergency, and a contingency plan is imperative for operating in an orderly manner.

Summary

The operating functions that support production activities have three primary characteristics:

◊ They are highly repetitive.
◊ They are processed at high speeds that can affect many accounts in a short time if something goes wrong.
◊ They involve no direct contact with the customer.

Card production is the function of embossing and encoding plastic cards. The specifications governing the physical positioning of embossed characters and encoded tracks are established by the national associations, which enforce their compliance. Card production must take place in a secure environment that protects against theft and abuse of blank and finished cards. Typically, the card production area is isolated, with access strictly limited to authorized personnel.

Work scheduling is critical. If accounts are unevenly distributed across billing cycles, the card production function faces concentrations of work on some days (or during some weeks or months) and idle capacity on others. Coordination of new account distribution and production schedule requirements is of major consideration to the bank's card production.

Sales draft processing involves the transfer of transaction information (merchant name and number, cardholder account number, and transaction date and amount) from the paper draft to an electronic medium (computer tape or disk). In some banks, this function is performed manually by keying the information into computer terminals. Other banks use automated equipment to capture the information from the paper draft at high speed. Preparing the drafts in working batches, repairing the drafts that were improperly keyed or read by the scanning equipment, making adjustments to merchant deposit accounts, and balancing are steps that must be completed before the transactions are sold to cardholder banks through interchange. Performance and quality assurance standards are key in this and in other production functions.

Statement production involves the printing and mailing of cardholder statements each month. Production time must recognize the federal regulatory requirements to mail statements to cardholders at least two weeks prior to the payment due date. Federal regulations also specify how finance calculations and charges are to appear on the cardholder statement. Proper editing and operating controls must be in place to make sure that the correct number of statements are printed and mailed each day.

Account distribution across billing cycles is a major concern in the statement production function. Even distribution provides the basis for an economic and efficient statement flow. Merchant statements must also be considered in statement production. Although federal regulations do not specifically govern the rendering of merchant statements, market competitiveness calls for timely and accurate production performance.

Remittance processing involves capturing and crediting payments for cardholder accounts. The receiving of remittances is important because cardholders include messages and returned cards with their remittance envelopes. Recording the date of payment receipt is critical so that—regardless of the processing date—the cardholder receives credit on the date the payment was received by the bank. The preparation of payments for batch processing, adjustments, and balancing must be completed before these payments are credited to cardholder accounts. Again, stringent performance and quality assurance standards ensure efficient operations.

The data processing systems that support the bankcard operation must also operate efficiently, by

◊ meeting processing windows
◊ performing daily file updates so that the operating departments at banks have current information to carry out their responsibilities
◊ maintaining response times so that the quality of customer service and production levels do not suffer
◊ maintaining the bank's computer system so that workflow proceeds uninterrupted

In addition, the bank must have proper audit controls to preserve the integrity of the processing systems and operating procedures.

Contingency planning is required by federal regulations for banks and should be in place for all departments of the bankcard operation. If an emergency does occur, backup systems and contingency procedures allow the operation to continue to function.

Review Questions

1. What are the three prominent characteristics of bankcard production functions?

2. Who establishes the specifications for the embossing and encoding of bankcards?

3. What are the basic steps in the processing of remittances?

4. Using the account expiration date, create a distribution of 55,600 new accounts across 20 billing cycles.

5. Why would a bank pay a third party to process its bankcard operation rather than doing it itself?

6. Suggest two ways to improve efficiency of the production functions.

8

COLLECTIONS

When an account fails to make a payment, the collection function begins. The collection staff plays the major role, determining how the delinquent customer should be treated and whether the situation is expected to be temporary or is likely to result in serious delinquency. Bank management strategies for both allocating collection resources and measuring productivity are important. Behavioral characteristics of delinquent customers need systematic monitoring and recoveries should be tracked and measured. A bank's use of automation can help the effectiveness of these collection efforts, but external influences, such as a regional recession, can thwart even the most capable collection staff. In addition, management must ensure compliance with the Fair Debt Collection Practices Act and monitor trends in delinquencies that may have long-term consequences to the bankcard portfolio.

After successfully completing this chapter, you will be able to

♦ explain how competition for payment affects collection

♦ list key characteristics of a successful collector

♦ discuss various measures of collector productivity

♦ explain how the process of behavioral scoring can improve the effectiveness of collection

♦ discuss the recovery process

Collection Pressures

Collection departments across the country have been under stress from forces on several fronts. Changes in federal bankruptcy laws in the early 1980s shifted much of the financial burden of overindebtedness from the borrower to the lender. The social stigma associated with declaring bankruptcy has been substantially softened. Some law firms, for example, regularly advertise their professional assistance in consumer bankruptcy proceedings. In some cases, the advertising message approaches the point of encouraging debt-burdened consumers to use this means to escape the pressures of indebtedness. Consequently, the collection staff cannot rely solely upon traditional debt obligation values and processes to obtain repayment.

Banks and other bankcard lenders have further aggravated the credit problem through their use of mass preapproved solicitations. Simply, many consumers have been given more credit than they can manage. Some regions in the United States depend heavily on industries that have recently faced hardship, such as farming and lumber. Consumers without jobs cannot pay their bills. All of these factors have placed an increased pressure upon the collection function in most banks.

Payment Competition

In one sense, collection is really a form of competition for payment. A cardholder with insufficient income to pay all of his or her bills will usually attempt to pay at least some of them. Many factors come into play in the borrower's decision about which lenders to pay. For example, the cardholder may hold a long-term relationship with a particular bank and feel a greater sense of responsibility to clear a debt with it. Or the cardholder may believe there is greater long-term risk to his or her credit rating if the bank remains unpaid. Even the collector's tone of voice, empathy, or willingness to work out an arrangement for repayment may influence whether the cardholder decides to repay the bank first. The reverse can also be true.

Most people do not borrow money with the expectation that they will not be able to pay it back. Even those that overextend their accounts generally believe that something—a windfall or salary increase—will allow them to repay their debt. Delinquency is most often brought about by an unexpected interruption of routine, such as the loss of a job, a severe illness, injury, or divorce. Whatever the cause, the bankcard collection function competes for the scarce resource of a cardholder's ability to repay. Consistency and professionalism can influence the cardholder at an early stage in the delinquency cycle, and the effective use of warnings is important in the competition for repayment.

What Makes a Collector Successful?

"I am not a collector. I am a facilitator," said a collector when asked what his job involved. "My job is to help people who are having financial difficulties figure out a way to repay us at least some portion of what they owe each month. Then when they get back on their feet, they can go back to a normal payment schedule."

Another person in the same department answered this way: "Some people will tell you the truth about why they're delinquent, even if it means they have to admit to having made bad judgments with the use of their bankcard. Others will tell you what they think you want to hear. My job is to decide which stories are genuine and treat them accordingly. A collector can't treat everybody the same way. Some people really want to cooperate and try to work out the problem. Some just don't care."

The dynamics of collection are such that not everyone can be effective. Although there is no common mold, several characteristics apply to collectors. They must be calm under pressure. They function constantly under the demands of management's expectations on one side and the tension of daily conversations with cardholders who are delinquent in payments on the other side. As you would expect, some cardholders become irate when repeatedly asked for money they do not have, or do not want to use, to pay their debts. Collectors must be discerning listeners and separate the genuine from the fake.

Collectors have to exercise good judgment. They have to recognize when delinquent balances can be collected and when the effort is fruitless—in some cases, even when the cardholder truly wants to repay but cannot. Collectors must be skillful negotiators and compete with other lenders for the scarce resource of the cardholder's money. As one collector said, "A little of something is better than all of nothing." Another collector makes the point this way: "I was working with 150 accounts of people in the farming business. A few years ago, the weather was bad and the harvest was low. These customers wanted to repay us, but they just didn't have enough money. They promised that they would pay us when the new crops came in. I believed them—they were honest people in trouble. So I asked each customer to send me at least $5 a month. The interest charges were waived because the accounts were severely delinquent. Each cardholder sent a check every month. The following year, all but two repaid their debts in full."

Management relies on the individual collector's judgment and negotiating skill. If management had interfered with the collector's arrangement with the customers in the last example, chances are the bank would have lost. If the collector's judgment had been unjustified, the bank would still have lost. Thus, the collector's task is to find the appropriate balance between being too easy

---

and being too hard on delinquent cardholders. Management's task is to find the balance between being too easy and being too hard on collectors.

The Fair Debt Collection Practices Act regulates collection activities and is specific in defining unacceptable collection practices. For example, cardholders cannot be telephoned by collectors before certain hours in the morning and after certain hours in the evening. Abusive language, threats, harassment of the cardholder at work, as well as discussion of personal financial information with anyone other than the cardholder, are prohibited.

The job of collection is diverse and demanding. The collector must exercise good judgment and understand when to be firm, when to assist or insist, when to expect payment, or when to write off the loan. In addition, the collector must distinguish between an isolated incident and a trend. A cardholder may have made just one mistake, or it could be the first in a series of problems that result in a loan loss.

Portfolio Monitoring

Monitoring the quality of receivables is key to keeping charge-offs low and can be accomplished through regular and systematic portfolio surveillance. Status reports—called delinquency account report, account aging report, or exceptions account report—are generally produced at least monthly for review by management (although some are produced daily for working with delinquent accounts). All accounts with balances are shown on the report, by number of accounts and dollar value, according to their account status. Exhibit 8.1 illustrates the information a typical delinquency account report will contain.

EXHIBIT 8.1   Delinquency Account Report (January 19XX)

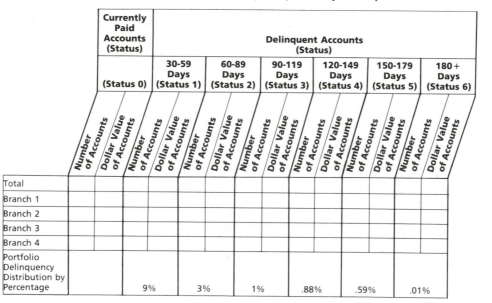

1. **Currently Paid Accounts (Status).** This section shows the number of currently paid accounts and their total outstanding balances. Accounts in which the last payment was posted (credited) on or before the payment due date are represented here.

2. **Delinquent Accounts (Status).** This section shows the number of delinquent accounts and their outstanding balances. Accounts are represented according to length of delinquency, that is, by the number of days after the payment due date the last payment was received. The further an account extends beyond the due date, the greater the likelihood it will be charged off. After 180 days, the outstanding balance is charged off to loan losses.

3. **Total.** This section shows the total number and total dollar value for accounts in each category.

4. **Branch Accounts.** This section shows the number of accounts and dollar value of outstanding balances for each bank branch, which will vary from branch to branch. The "total bank" should always be subdivided on some systematic basis: by branch, geographic region, source of solicitation, etc.

5. **Distribution of Portfolio Delinquency by Percentage.** This section shows the distribution of delinquency in the total bankcard portfolio. In a mature portfolio with sound credit quality, the vast majority of accounts should be current. After the first cycle of delinquency (30 to 59 days past due), the level should decline dramatically as collection efforts are made to resolve delinquent accounts (see exhibit 8.2). Some cardholders in status 1 have simply forgotten to make their most recent payment—a delinquency more academic than real. Nonetheless, the report must present the account according to its actual status.

EXHIBIT 8.2   Impact of Collection Effort on Delinquencies

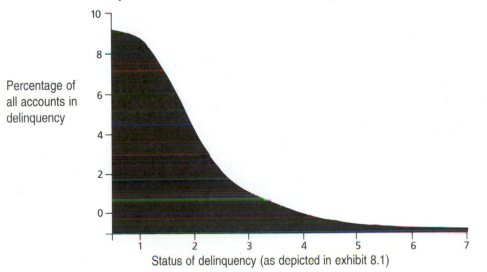

Percentage of all accounts in delinquency

Status of delinquency (as depicted in exhibit 8.1)

Systematic portfolio surveillance is critical to monitoring the overall quality of the portfolio and managing the collection effort. Depending on circumstances, such as a regional recession or the declining effectiveness of the collection function, portfolio quality can quickly deteriorate unless it is strictly monitored.

Collection Tools

Beyond the obvious telephone and clerical operations, information and authority represent two important categories of support mechanisms. First, the collection function requires as much information about a delinquent cardholder as possible. A telephone number and address where the cardholder can be reached is most important. This may not be as easy as it seems. Often, people who cannot pay their bills have their telephones disconnected or move to another address. The collector may have to call the last known place of employment to see if the cardholder still works there or try to obtain the name of a relative or friend for more information.

If the cardholder has left the area altogether to avoid collection attempts, the collector may turn the account over to a skip tracer. This individual uses a network of contacts (social services, law enforcement agencies, other lenders, relatives) to track down the delinquent cardholder.

skip tracer
A person whose job is to find delinquent cardholders who have vanished to avoid collection attempts and repayment of monies owed.

Sometimes the delinquent cardholder simply will ask someone else to screen telephone calls. If the collector is calling, the cardholder is never there. The collector then must find ways to reach the person, even if this means leaving several messages at work or with friends. However, the collector *cannot* divulge any financial information about the delinquency to anyone but the cardholder. The collector may only leave messages requesting a return call to the bank. A lot of persistence is required when a cardholder does not want to talk.

Credit history with the bank and other lenders is also revealing. An experienced collector can review a cardholder's credit history and determine whether a particular instance of delinquency is an isolated event, limited to the bank, or the result of overindebtedness affecting other lenders as well. Difficulties with other lenders indicate a serious problem and reduce the likelihood of quick resolution to a bank's own delinquent cardholder account. Similarly, a high proportion of recent cash advance activity may indicate that the cardholder is having financial problems and is drawing cash before the account is closed. This information will determine the course of action the collector will follow. If the account appears to be collectible, efforts to do so will be undertaken. On the other hand, if collection possibilities are limited, the collector may charge off the account and assign it to a recovery function, which may be a department within the bank or an outside collection agency. In the case of serious delinquency, the collector should attempt to retrieve the card, or cards, from the cardholder. Continued use of the card increases the bank's exposure to loss.

recovery function
The activity involved in trying to collect the balance owed after an account has been charged off.

Documentation of collection experience for each account is critical. If a cardholder keeps commitments, or fails to keep them, the file should reflect that behavior. Agreements between the collector and cardholder (repayment amounts each month, payment due dates, and so on) must be documented. Telephone numbers, names of people who may know how to contact the cardholder, and other relevant information also must be documented and kept on file. If the cardholder becomes delinquent again, or if another collector takes over the account, good documentation expedites the collection effort. More often than not, documentation makes the difference between dollars collected and dollars charged off.

The second kind of collection mechanism, authority, concerns decisions about what happens to delinquent accounts. A collector may decide early on that an account is uncollectible and send it for charge-off. On the other hand, the collector may negotiate an arrangement for repayment that is outside the bank's normal policy. For example, the collector may agree to payments of $5 per month when the normal schedule calls for a minimum of $25 per month. The collector may decide to re-age an account (bring the account back to current status through an entry to the file) if the cardholder is keeping agreed commitments. The authority granted to the individual collector must be based on his or her experience and performance. This authority also must be clearly defined in policies and procedures for the collection function and the collector's performance carefully monitored by management.

Productivity Measures

The most obvious measure of collection effectiveness is the amount of dollars collected. But the amount of total dollars collected in relation to total delinquent dollars is affected by a number of factors. Let's consider some briefly.

♦ **The number of accounts assigned per collector** is based on ability and experience, on the information available on accounts (fewer assignments can be worked if the collector has to develop information on the cardholder), and on account collectibility (advanced cases of delinquency take more time).

♦ **The number of accounts a collector works each day, week, and month** is important to monitor. A relatively low number of accounts worked could indicate that time and effort are being wasted. A high number could mean that the collection effort is superficial, with too little time spent on each account.

♦ **The number of accounts charged off** is a key measure. Too many accounts passed on for charge-off may mean that the collector is not performing well or that an external influence, such as rising unemployment, may be starting a trend. Thus, management must carefully monitor unexpected plant layoffs, for instance, and other economic conditions.

♦ **The number of re-agings** must be carefully monitored. The effect of re-aging is to hold delinquency at its present level or reduce the level by bringing accounts to a current status. Unwarranted re-agings make the collector look good on the surface because accounts are "cured" of delinquency. However, re-aging understates the real level of delinquency and exposure to loss.

Automation

A bank's degree of automation influences the level of its collection productivity. In a manual operation, the collector has to dial the call (and try repeatedly if the line is busy or no one answers), write out the details of the conversation with the cardholder, draft a letter or give instructions to support staff, file and retrieve documentation manually, and so forth. While these activities have to be performed, they reduce the amount of time the collector can actually spend in conversations with cardholders.

Highly automated operations, on the other hand, expedite these activities. Let's visit the automated collections department of ABC Bank.

John, one of the members of the collection staff, has just arrived at his desk. John works through the afternoon and evening hours because ABC Bank runs shifts from 8:00 a.m. to 9:00 p.m. to attain the most coverage. (There are Saturday shifts as well.)

John puts on a headset and enters his personal code into the terminal. The first account he will contact, Cardholder A, appears on the screen.

A brief history is displayed, showing the name of the cardholder, the date of the last payment, and highlights of prior discussions other collectors have held with the customer. After a review of the information, John enters the "call" command by pushing the appropriate key on the terminal.

The system automatically places the call and, in just a few seconds, John hears a busy signal. The information on Cardholder A disappears from the screen and is immediately replaced by information for Cardholder B. The system puts Cardholder A in a hold status to call again later.

John reviews Cardholder B's data and pushes the call command. This time, someone answers, but the person on the other end of the line is not Cardholder B. John leaves a message for Cardholder B to return his call as soon as possible. He then enters a code that tells the system the present date and time and types in the name of the person who took the message. Finally, John enters a finish code and the information for Cardholder B disappears from the screen. Immediately, information for Cardholder C appears. John reviews the screen and pushes the call command. After several rings and no answer, the information for Cardholder C disappears from the screen. The system stores Cardholder C's data to try again in about one hour.

As soon as the screen clears, the data on Cardholder A reappears. The system places another call and this time the line is not busy. Cardholder A answers. During the conversation, John enters a "promise to pay" code. A portion of the screen displays the following message:

Cardholder A will pay $ on   / / .

John enters the amount and the date behind the prompts. He concludes the conversation and enters the finish code. The screen clears and displays Cardholder D's data. Cardholder D promised to make a payment three days ago, but it has not been received. Cardholder D answers and tells John that a payment will be made today at the bank. John enters the information on the terminal. The ABC Bank's system has been preprogrammed for payments received to record the cardholder's account status as

◊   promise kept (re-age and bring current)
◊   maintain present status until three successive payments are made
◊   change to another status

John will not see the account again unless additional collection effort is needed. After John pushes the finish code, Cardholder E's data appears on the screen. The system places the call, which is answered. The person tells John that Cardholder E has moved to a new address but has no telephone. John enters another code and the following message appears on the screen:

New address: _____

New telephone number: (   ) -

John enters the new address and enters a "not available" code after the prompt for the new telephone number. The system then asks John if he wants to send a letter and he responds affirmatively.

Send letter: 1   Contact date: 23

John enters the choice for form letter 1, which tells the cardholder to contact him between the hours of 1:00 p.m. and 9:00 p.m. on or before a specified date. The system automatically prints out form letter 1 and addresses it to Cardholder E from John. If no response is made, the system will bring the account to John's attention on the day after contact is due. John also enters the name of the person who gave the new address for Cardholder E in the memo portion of the screen and pushes the finish code. This procedure continues until John leaves for the day.

This high degree of automation allows John to spend more time working on accounts because he is not required to spend time on support activities. Banks that use automated systems establish productivity measures that reflect the expectation of higher collector productivity.

Behavioral Scoring

Some banks use a process of behavioral scoring to help improve collection effectiveness. The purpose of behavioral scoring in collections is to predict the likelihood of an account's continued delinquency or collectibility. Like credit scoring, the behavioral technique uses profiles as the reference points for predictability.

behavioral score
A score that predicts the account behavior most likely to occur according to statuses ranging from current to charge-off.

The behavioral scoring technique is dynamic, in that information on the account is kept current. For example, payment history, balance fluctuation levels, balance in relation to credit line, changes in job or area of residence, and similar information are maintained. Every relevant transaction or change is immediately added to the account history, which is later reflected in cardholder profiles. This history becomes the basis for assigning behavioral scores to accounts in the collection stream. The use of these scores becomes an inherent part of a bank's collection strategy. Let's consider a simple example.

Exhibit 8.3 illustrates a behavioral score on a delinquency continuum. In this example, the 0 to 100 range represents the spread from very mild delinquency to charge-off.

EXHIBIT 8.3   Behavioral Scoring on a Delinquency Continuum

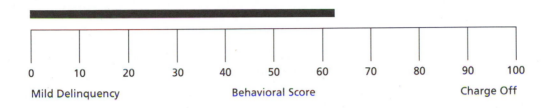

A behavioral score of 0 to 10 indicates that cardholder payments are occasionally late but are not expected to progress to serious stages of delinquency or charge-off. A reminder notice that the account is past due is all that is required. There is no need for a collector to make direct contact.

A behavioral score of 10 to 20 indicates that cardholder payments are frequently delinquent for a short time. However, these accounts will become current before advancing to a serious stage on the continuum. A reminder notice to the cardholder is adequate and no direct collector contact is required.

A behavioral score of 20 to 30 indicates that cardholder payments are in early delinquency cycles and approaching the credit limit. The cardholders of these accounts require a reminder notice and direct contact with collectors to avoid advancing along the delinquency continuum. Progressive behavioral scores indicate additional action on the part of the collector.

However, a behavioral score of 90 to 100 indicates that accounts will not be paid and are to be charged off immediately. These accounts are assigned to a collection agency. At this point, the use of reminder notices and collectors' efforts are ineffective and a wasted expense.

Behavioral scoring allows management to stay informed about the allocation of collection resources. In this example, neither accounts scoring from 0 to 20 nor accounts scoring from 90 to 100 would be assigned to the collection staff. Depending on the way the bank constructs its delinquency continuum, less experienced collectors would work accounts with lower behavioral scores, while more experienced collectors would work the accounts with higher behavioral scores. The major consideration is that behavioral scoring helps management assess collectibility and allocate collection resources accordingly. This leads us to consider some bank assignment strategies.

Collection Assignment Strategies

Collection strategies are many and vary from bank to bank. The following paragraphs give you a sense of just some of the different strategies that can be employed.

♦ **Assignment by Delinquency Cycle.** This strategy simply involves the assignment of accounts to collectors according to the status of delinquency. Some banks assign accounts at 30 days past the payment due date. Others assign accounts at 40 or 60 days. This strategy may also be employed according to the manner in which the account was originally acquired. For instance, accounts acquired through a preapproved solicitation could be assigned at 15 or 30 days of delinquency. On the other hand, accounts that were approved on the basis of a complete application could be assigned at 45 or 60 days of delinquency.

♦ **Assignment by Collector's Experience.** Under this strategy, less experienced collectors work accounts that are more easily collectible. Experienced collectors work accounts that are more difficult.

♦ **Assignment by Account Balance.** Here, accounts are stratified, or divided, by the amount of outstanding balance. Regardless of the stage in the delinquency cycle, accounts are distributed to collectors according to the balance.

♦ **Assignment by Geography.** This strategy assumes that collectors' familiarity with different geographical areas will facilitate the effectiveness of collection efforts.

♦ **Rotating Assignments.** This strategy involves account groups that are periodically rotated to different collectors. The assumption is that a new person contacting the delinquent customer will facilitate collection.

♦ **Assignment by Behavioral Score.** This strategy calls for assignment of accounts based upon predicted collectibility and the collector's ability to persuade the cardholder to repay a debt.

A highly automated operation lends itself to experimentation with various strategies. More information can be accumulated and analyzed in a shorter period of time to measure the effectiveness of these methods. In the example of automation at ABC Bank, management has a predetermined strategy for assigning accounts to John. The assignment criteria determine the specific accounts assigned and the sequence in which they will be worked. The strategy also calls for the actions John takes in entering codes and other relevant information, as well as the intervals between telephone contacts and other follow-up actions.

The use of automation and behavioral scoring offer prospects for developing more effective collection strategies in the years ahead as changing economic circumstances, new lending practices, and new consumer habits force banks to reevaluate their collection practices.

Recoveries

After an account has been charged off, it is transferred to a recovery mode. The approach to collecting the account changes, as well as the accounting classification on the bank's financial statements. Many banks use outside collection agencies to attempt to recover some portion of the balance charged off. Some use an internal recovery department. Some banks use both. Legal recourse, too, is often used in recovery activities, such as small claims courts, attachment of assets, and claims on proceeds from the sale of assets.

The effectiveness of recoveries must be measured. Banks frequently calculate the total amount of dollars recovered as a percentage of total dollars charged off. Given that a time lag exists between charge-off and recovery (6 to 12 months is typical), the following examples indicate how recoveries may be measured:

♦ Recoveries in the last six months of a year divided by total charge-offs in the first six months of the year equal the percentage of recoveries.

♦ Recoveries in the first six months of a year divided by total charge-offs in the last six months of the previous year equal the percentage of recoveries.

Thus, a monthly calculation should be made that recognizes the time differential between charge-off and recovery.

Comparisons of recoveries with other banks or collection agencies are also used. But the comparisons must correlate. For example, a comparison of recoveries between a large bank in Los Angeles and a small bank in Omaha would be questionable. Not only do the two cities differ in size, they also differ in population (and perhaps in social attitudes). In addition, any significant difference in lending practices at the two banks will influence recoveries.

Management Concerns

Management concerns fall into the following areas:

♦ **Collection Effectiveness.** This involves the total dollars in delinquency and charged off as a percentage of total amounts outstanding.

♦ **Collection Efficiency.** This involves the number of collectors and support staff required in the collection function, which is expressed as the cost of the collection resource. For example, 30 cents of collection expense for each dollar collected is obviously better than 60 cents of collection expense for each dollar collected.

♦ **Recovery Levels.** These center on an individual bank's performance and how that compares with correlating banks.

♦ **Charge-Off Trends.** These concern the direction of charge-off levels and how current levels compare with those of prior years and with the current experience of similar banks.

♦ **Emerging Trends.** These concerns look at bankruptcy-related charge-offs, regional delinquency and charge-off levels, changes resulting from modifications to approval criteria implemented in the past year, and the like.

Fundamentally, management is concerned with the overall quality of the bankcard portfolio: where the level of risk (as measured by delinquency and charge-off levels) has been, where it is now, and where it is headed.

Interaction between Credit and Collections

Good communication between the credit approval and the collection staff is critical. For the most part, the collection function is the recipient of actions that take place in the credit function—changing approval criteria, marketing programs to acquire new accounts, and so on. Changes at the front end inevitably ripple through to collections. Thus, it is critical for the credit staff to communicate with the collection staff. Even a moderate increase in the number of new accounts will eventually affect resource requirements for the collection function.

Information must flow from the collection staff to the credit staff, too. While the collection department is the first to see the consequences of changes in approval criteria, it is also the first to see geographic trends or industry-related problems associated with layoffs or reductions in force. Account behavior characteristics (such as high levels of cash advance activity or balances exceeding the credit limit) need to be fed back to the credit staff.

Some banks hold formal and regular meetings to ensure that the needed interaction between credit and collection personnel occurs. Other banks are much less formal but insist that some form of dialogue occur. Whatever the means, what is important is the timely, relevant, and complete exchange of information between the two staffs. Reviewing a complete credit file after a charge-off, for example, may be a valuable learning experience for a loan officer making credit decisions. This is one of the most important ways to strengthen the overall lending and collection process, while preserving the quality of the bankcard portfolio.

Summary

When accounts become delinquent, the collection function takes over. The objective, of course, is to bring overdue accounts into a current repayment status. Some banks use a method called behavioral scoring to help predict the likelihood of account collectibility so that collection efforts can then be more efficiently deployed. To be most effective, however, the credit and collection staffs must have a formal and regular exchange of information. This exchange improves the overall effectiveness of their respective functions. Prudence in lending money will decrease the burdens of the collection function. Consequently, the level of risk and overall profitability for the bankcard portfolio will be in balance.

Review Questions

1. How does competition for payment affect collections?

2. What is meant by collector negotiation?

3. What risk exists with collector re-agings?

4. What is the purpose of behavioral scoring in the collection function?

5. Identify ways to ensure an appropriate level of interaction between credit and collection staffs in a bankcard operation.

6. Create a simplified model for collection resource allocation based on behavioral scoring results.

9

ADMINISTRATIVE SUPPORT

The production functions discussed in chapter 7 are essential, but the show could not go on without the support functions: accounting activities, exception item processing, and administrative services.

Depending on the individual bank's organizational structure, some activities performed in the accounting support area may be assigned to either the bankcard accounting department or the operations department. Where they are assigned is not important; how they are performed is. The designation "accounting support" refers largely to the direct interface with the bankcard general ledger (and ultimately the bank's overall general ledger) and to the financial interface with the MasterCard International and Visa International settlement systems.

We also look at the nature of adjustment activities resulting from disputed transactions.

After successfully completing this chapter, you will be able to

♦ identify and describe the two major accounting support functions

♦ discuss the major activity of the exception processing function

♦ explain the difference between the two types of chargebacks

Accounting Support

Accounting support involves direct interface with the bankcard general ledger (and ultimately with the bank's general ledger) and financial interface with the MasterCard and Visa settlement systems. The two functional areas of accounting support are adjustments and settlement.

Adjustments

The adjustment function plays a significant role in the bankcard operation. It holds little glamour, but keeps the general ledger in balance and is the means for correcting customer accounts. Adjustment activity involves both merchants and cardholders and is another component in the formula for providing a competitive level of customer service.

The major activities performed in the adjustments area include merchant adjustments, payment adjustments, ATM transactions, returned checks, and, in some banks, the bankcard checks that function as ordinary personal checks but actually draw funds from the bankcard line of credit. Brief descriptions of these activities follow.

♦ **Merchant Adjustments.** These involve the review, correction, and resubmission of entries resulting from processing drafts and merchant deposits that are rejected for invalid or missing account numbers. Adjustments may also be made to deposit amounts that were incorrectly entered as a result of errors on the original merchant deposit.

♦ **Payment Adjustments.** These involve researching and correcting posting errors and unposted payments largely received from the customer service department. The adjustment group also researches and resolves adjustments received from the data processing center for missing items or items that cannot be associated directly with a specific account.

♦ **Returned Check Handling.** This involves both resubmitting or debiting cardholder accounts for checks returned for nonsufficient funds (NSF).

nonsufficient funds
The designation given to checks that are returned when there is not enough money in the checkholder's account to pay the check when it is presented to the bank for payment.

♦ **ATM Adjustments.** These involve researching and processing problems with ATM cash advance transactions.

The adjustment function interfaces with many other functions in the bankcard operation, such as sales draft and payment processing, collections, customer service, returned items, and settlement.

Performance and Quality Assurance Standards

♦ **Merchant Adjustment Timeliness.** Timeliness is a concern from the standpoints of operating effectiveness and customer service. Keeping adjustment entries current is necessary to control the bank's exposure to loss. Making sure that the merchant's account reflects current activity is obviously of concern to the merchant. This standard states the average number of days from receipt of the adjustment journal (listing the items to be worked) to the date that all adjustments on the journal are cleared. A sample adjustment journal appears in exhibit 9.1.

EXHIBIT 9.1   Unposted Monetary Items

Cardholder Account Number	Merchant Account Number	Reference Number	Authorization Number	Transaction Date	Transaction Code	Reason not Posted	Amount	Last Transaction Date

This journal provides the adjustment staff with reasonably complete information with which to process the adjustment. As you can see, both cardholder and merchant account numbers appear, although this report is only for cardholder adjustments.

♦ **Merchant Adjustment Accuracy.** This standard integrates speed with the quality of the work performed. One way to express accuracy is as a percentage of total transactions processed, such as x percent of the adjustments processed will be accurate the first time they are handled.

♦ **Payment Adjustment Timeliness.** The timely correction of errors is essential in diffusing customer dissatisfaction. Therefore, the standard is stated as the average number of days from receipt of the payment adjustment request to the date that the payment discrepancy is resolved.

♦ **Payment Adjustment Accuracy.** This standard integrates speed and quality of performance.

♦ **Suspense Processing Timeliness.** In many departments of a bankcard operation, some items appear daily but cannot be immediately processed. They are then moved to a suspense account until the problems can be resolved. Suspense accounts handle a great deal of transaction activity. Every item that is entered into suspense eventually has to be resolved and taken out of suspense. The timeliness of clearing transactions is a major concern to management because older items are much more difficult to resolve. This standard states the average number of days from the date the item is entered into a suspense account until the date the item is cleared. A companion standard, calling for accuracy of suspense processing, preserves performance quality.

Settlement

Settlement is the means by which the bank sells its merchant draft transaction values to MasterCard and Visa. Settlement also enables the bank to buy its cardholder transaction values from these associations. Every working day, the bank settles with the networks; for example,

Bank sells merchant draft values	$1,000,000
Bank buys cardholder transaction values	$ 500,000
Net settlement (assoc. pays bank)	$ 500,000

or

Bank sells merchant draft values	$ 500,000
Bank buys cardholder transaction values	$1,000,000
Net settlement (bank pays assoc.)	$ 500,000

These kinds of calculations are performed daily in the settlement area. Exhibit 9.2 illustrates the workflow that settlement performs.

EXHIBIT 9.2   Settlement Process

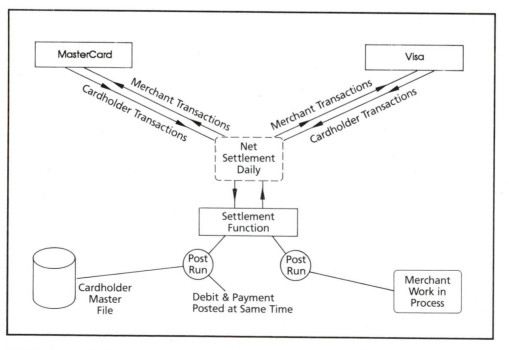

Settlement, though invisible to cardholders and merchants, is a critical part of the bankcard operation. It enables merchant and cardholder transactions to be exchanged; it accounts for every dollar of value and, indeed, every transaction that occurs between merchants and cardholders. This function is part of the foundation of the bankcard operating structure and requires a high level of management emphasis.

Performance and Quality Assurance Standards

Clearly, performance level is a matter of significant emphasis by management. Several standards apply:

- **Settlement Timeliness.** Not surprisingly, this standard calls for settlement to be completed each working day by a deadline established by each association.

- **Settlement Accuracy.** Equally important is the correctness of the settlement: 100 percent accuracy in preparation and clearance of settlement entries should be mandatory.

- **Unposted Item Resolution.** Some items will not be properly posted during the course of processing. They must be accounted for and posted correctly. Standards for promptness and accuracy must meet management's expectations. The timeliness standard sets forth the average number of days from receipt of the unposted item report to the date the item is cleared. The accuracy standard sets the expected level of accuracy in the resolution of unposted items.

- **Account Certification Frequency.** This standard deals with both general ledger accounts and demand deposit accounts. The interface with the MasterCard and Visa networks is accomplished electronically through demand deposit accounts. Because of the number of transactions flowing into and out of these accounts, the possibility of error is high. The process of certification ensures that errors have been identified and corrected. Each bank establishes the frequency of certification on a daily, weekly, monthly, or quarterly basis, depending upon the nature and volume of the accounts and the level of exposure involved.

- **Suspense Processing Timeliness.** The settlement function also has suspense accounts for holding items until they can be correctly entered into the system. As with other suspense accounts, older items are generally more difficult to clear. This standard establishes the average number of days from the date the item is placed into a suspense account until the date the item is resolved and cleared from the account. A companion standard for accuracy integrates speed and quality of the work performed.

Several other accounting support functions should be mentioned: refunding credit balances on cardholder accounts, handling stop payment requests, and charging off entries to the general ledger. Responsibility for these activities is determined on an individual bank basis, so they reside in various organizational units in different banks.

The same is true of adjustments and settlement. They may appear in the finance division in one bank and the operations division in another. The main consideration is that they are carried out efficiently and accurately. Management of the bankcard operation has to establish specific standards that define efficiency and accuracy to ensure acceptable operating proficiency and good customer service. Management also has to consider internal operating controls and segregation of duties.

Exception Processing

The process of handling transactions that must be treated as exceptions, or chargebacks, is specific and must be followed precisely in accordance with regulations established by Visa and MasterCard (which may vary between the two) and within strict timeframes.

A chargeback is technically initiated by the cardholder bank on behalf of the cardholder. In some cases, the cardholder is disputing the transaction; in others, the bank initiates the chargeback without the cardholder's involvement. The merchant bank can present the item again to the cardholder bank when additional information is supplied, when there is no documentation to support the chargeback, or when the cardholder bank chargeback is incorrect. When chargebacks and re-presentments enter the bankcard operation, there are three alternatives:

- ◊ defend the merchant or cardholder and return the item or transaction to the bank disputing it and achieve resolution there
- ◊ charge the item to the merchant or cardholder account
- ◊ absorb the amount of the item in question as a loss

In virtually all cases, the first alternative is the most desirable one. The third alternative is used as a last resort. Both national associations offer a fee-based service to speed up the retrieval and chargeback resolution process.

Staffing the Exception Function

For our discussion, assume that the following areas handle specific segments of the exception management process: general merchant, re-presentments, and major merchant sections. The process is essentially one of defending the position of the cardholder or merchant.

- ♦ **Merchant Area.** This area reviews and processes incoming chargebacks against the bank's merchants. The staff first examines the chargeback to make sure that it appears to be valid. Most of the time, this preliminary review is followed by research into the reason for the chargeback in order to protect the merchant against unwarranted chargebacks.

  Bank policy may be to simply pass the chargeback (and information associated with it) to the merchant. The merchant can then perform whatever research is necessary for defense against it. If the transaction is to be charged back or re-presented to the cardholder bank, the staff makes sure that MasterCard International and Visa International regulations are followed. The regulations define when it is appropriate to

  - ◊ re-present the item to the cardholder bank
  - ◊ pass the amount on to the merchant for charge-off
  - ◊ pursue compliance or arbitration with the MasterCard or Visa organizations

compliance/arbitration
A resolution procedure that takes place when the cardholder bank and merchant bank cannot arrive at an agreement on a chargeback.

  MasterCard and Visa have established precise rules that govern the chargeback process. The procedure is technical and the loser of the case is subject to an assessment by the association. (The assessment is intended to encourage banks to work out their differences.) The resolution step is referred to as compliance by MasterCard International and arbitration by Visa International.

  The merchant area also prepares accounting entries resulting from chargeback activities. These might involve determining whether to charge the merchant, re-present the item to the cardholder bank through interchange, or absorb the loss at the bank (usually, this occurs only when the bank is in error because of mishandling the chargeback).

- ♦ **Re-Presentment Section.** A re-presentment occurs when a merchant bank believes the cardholder or cardholder bank unjustifiably initiated a chargeback against the merchant bank. The item is returned to the cardholder bank to charge the cardholder's account again. This area reviews and processes the item being re-presented. It is usually necessary to research the issue further, which includes direct conversation with the cardholder. Compliance with MasterCard International and Visa

International regulations must also be ensured. The re-presentment staff has three alternatives:

   ◊   charge the cardholder's account
   ◊   charge the item back to the merchant bank
   ◊   under circumstances of bank error, absorb the loss

♦ **Major Merchant Area.** Large merchant customers may generate sufficient activity that a bank will establish a department dedicated to this group. If a merchant asks the bank to process a re-presentment, the bank ages the items so that regulated timeframes are not exceeded. Items are entered into suspense accounts until the merchant has researched the chargeback and decided either to absorb the chargeback or to proceed with the next chargeback step and re-present the item to the cardholder bank.

The exception management area can be highly volatile. Proper management controls must be in place to avoid losses. The national associations' timeframes are absolute. If missed, there is no negotiation—the bank in error takes the loss. Items or transactions in the exception management queue must be monitored daily by management using a chart similar to that in exhibit 9.3.

EXHIBIT 9.3   Summary of Daily Chargeback Activity
(Date)

Receiving ICA/BIN XXXXXXXXXXX		Page 1
Item	**Count**	**Gross Amount (Dollars)**
Starting suspense totals	31	2,395.66
Cleared details		0.00
Reverse charge back	0	0.00
Charge off	0	0.00
Cardholder adjustments	0	0.00
Merchant adjustments	0	0.00
Cleared total	0	0.00
New entries		
Manual items	0	0.00
Transmitted items	1	90.00
Draft corrections	0	0.00
New entries total	1	90.00
Closing suspense totals	32	2,485.66
Aging control report		
0 through 15 days	2	123.15
16 through 30 days	1	142.00
31 through 60 days	25	2,127.84
61 through 90 days	3	75.16
91 through 120 days	1	17.51
Compliance/arbitration	0	0.00
Grand total	32	2,485.66

If you look at the exception report, you can see that all entries are accounted for in the Cleared Details and New Entries sections. The aging depicts the status of individual items held in suspense pending resolution. Management is concerned that the total amount in suspense is reasonable within the operating framework of the bank and that the distribution of the aging makes sense. The total amount in suspense will vary depending on both the activity and performance levels of the individual bank. Agings will vary depending on the stages that items occupy in the chargeback cycle, as well as on the performance level of the function.

Performance and Quality Assurance Standards

♦ **Outgoing Chargeback Processing.** Outgoing chargebacks originate when a cardholder challenges a transaction. This standard sets forth the average number of days from the receipt of the cardholder's challenge (by telephone or letter) to the date the transaction is charged back to the merchant bank.

♦ **Outgoing Chargeback Accuracy.** Documenting the reason for the chargeback is an important part of the process. This standard calls for complete documentation to support the chargeback. If the chargeback is incomplete, it can be returned to the cardholder bank as a re-presentment resulting from insufficient documentation. Time is lost and unnecessary effort expended. As a result, the standard may require 99 percent of properly documented chargebacks on a continuing basis.

♦ **Incoming Chargeback Processing.** When a chargeback is received, it must be entered into the appropriate suspense account until it can be resolved and cleared. This standard establishes the average number of days from the receipt of the chargeback to the date it is cleared from suspense. This standard may state the expected performance level as follows:

　◊　All items must be cleared to meet MasterCard and Visa standards.
　◊　90 percent of the items will be cleared within 20 days.
　◊　100 percent of the items will be cleared within 30 days.

A companion standard that establishes the level of accuracy that management expects will preserve quality in the function.

♦ **Suspense Processing Timeliness.** The timely processing of items in suspense is critical. Older items are more vulnerable to charge-off. This standard states the average number of days from the date an item is placed into a suspense account until it is cleared. A companion standard that calls for the correctness and balancing of the suspense account is required to integrate speed and quality.

Exception items will also be charged back for reasons other than a cardholder challenge. These include chargebacks for transactions on expired cards, transactions on accounts that were closed prior to the transaction, and the like. Regardless of the specific reason for a chargeback, the process must be carefully executed to protect the bank from exposure to loss.

Although exceptions fall outside the routine processing stream, the volume of items falling into the exception category can be substantial. Because the regulations governing chargebacks are explicit, management must ensure that performance levels are adequate and timeframes are met. Performance level is a function of

- ◊ trained and adequate staff
- ◊ operating and accounting controls
- ◊ explicit performance and quality assurance standards

Administrative Services

The overall bankcard operation, regardless of size, requires administrative support on several fronts. Data entry for file maintenance, telecommunications (to support collections, customer service, and authorizations), facilities maintenance, and word processing are included in administrative services. Larger operations generally create units whose responsibilities are to serve the entire operation.

Data Entry

The data entry function is responsible for all monetary and nonmonetary entries to the masterfile. Some banks distribute file maintenance among several departments. Other banks centralize the function in one unit. Regardless of the organizational arrangement, some standards are appropriate for high performance and quality assurance preservation.

♦ **Monetary Entry Timeliness.** Monetary entries include finance charge adjustments, merchant deposit adjustments, and cardholder credits and debits—any entry that involves money. Making sure that the entries are processed in a timely manner is of continuing concern to management. This standard states the average number of days from receipt of the monetary data to the day the entry is made. Generally, the timeframes set by this standard are short—one day from receipt is typical.

♦ **Monetary Entry Accuracy.** This standard establishes the percentage of total entries processed daily, weekly, and monthly that are accurate.

Incorrect entries cause customer dissatisfaction and create exposure to loss, so the required level of accuracy is normally very high.

♦ **Nonmonetary Entries.** These entries include name and address changes, account status changes (such as delinquency status), and other changes to the masterfile that do not involve money. Standards for timeliness and accuracy are necessary for high efficiency and quality levels.

♦ **Suspense Processing.** Data entry, too, has transactions that have to be held until a problem can be resolved. Standards pertaining to speed and accuracy are necessary here, as they are for all suspense account activity. Items that have aged a long time often end up as charge-offs.

Telecommunications

The focus as we discuss telecommunications will be on three areas of management concern: telecommunications design, capacity, and contingency planning. Underlying all of these areas is the concern for cost and benefit.

No single design fits every bankcard operation. On the contrary, there can be as many designs as there are bankcard organizations. Design refers to the particular kind of telecommunications equipment and how it is configured to meet the bank's operating requirements. Most bankcard telecommunications systems are part of the bank's overall system. Large bankcard operations have a dedicated staff to handle internal and external communications needs.

The major concerns of management are that

◊ internal communications be efficient
◊ customers can call into the bankcard area with relative ease
◊ outgoing calls are as inexpensive as possible

Management's concern with capacity is that the system is able to handle current levels of activity and provide some capacity for growth. Too much capacity is expensive and wasteful. Too little will hinder employees and annoy customers. For example, if the incoming telephone lines are inadequate for the volume of calls, customers will frequently get busy signals when they attempt to call. Inadequate outgoing capacity will inhibit the bank's productivity by causing employees to sit and wait for an outgoing line to become available. The challenge is to balance resources with needs. Most banks depend on experts to analyze their capacity requirements and install the appropriate system to provide the capacity they need. A large volume of incoming and outgoing calls will justify the added expense for greater sophistication of equipment and systems.

Contingency planning is important for a smooth operation. If computer and telephone lines are down, alternative paths are necessary until the condition is remedied. For example, an automated call distributor (ACD) system can be used to illustrate how contingency planning can be carried out. Exhibit 9.4 shows how the ABC Bank has two ACD systems (because of its large volume of incoming calls).

EXHIBIT 9.4   ABC Bank Automated Call Distributors

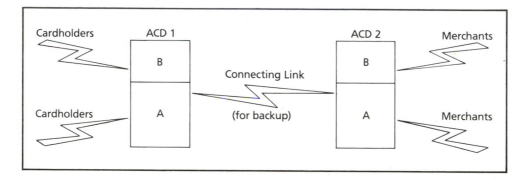

The two systems are linked and provide backup for each other when needed. ACD 1 is for calls coming into the customer service area. ACD 2 is for calls coming into the authorization area. In each case, the incoming call is routed to an available representative on a first-in, first-out basis. If all representatives are busy, a recording acknowledges the incoming call and places the call in queue. Each ACD in the illustration is divided into two parts. Part A shows the imaginary usage level in each ACD, while part B shows its idle capacity.

If the first ACD fails, all incoming calls are immediately diverted to the second ACD. Although there is insufficient idle capacity to absorb all the calls rerouted from the first ACD, a large percentage of calls can be taken until it is restored. The assumption is that the failure will last for only a few minutes, or at worst a few hours. Calls that cannot get into the backup ACD will receive a busy signal and have to keep trying. But the ability to receive a large percentage of calls from the failed ACD is more desirable than not being able to receive any. As exhibit 9.4 shows, management has designated the ability of each ACD to accommodate about 140 percent of current volume. If one ACD temporarily fails, about 40 percent of calls can be handled by the other ACD. Greater capacity is an added expense, so that having each ACD be able to handle twice the present level of activity is not common.

Having an alternative telecommunications system is as important as having a contingency plan for data processing. The difference lies in bank management's expectation of the magnitude of the problem. Situations will occur and must be planned for, so that the loss of a line or a temporary equipment failure will not seriously affect operations.

Facilities Maintenance

The facilities maintenance function may be a part of the bankcard operation or assigned to another unit in the bank. This function is concerned with the operating efficiency of equipment, for example, CRTs and incoming power lines. However, it also deals with other areas that can directly affect employee productivity, such as temperature, lighting, office furniture, noise levels, and air quality. It is sometimes easy to take this function for granted. However, employee health problems can range from neck strain to stress associated with long-term use of CRTs if the facility is not properly furnished and maintained.

Other Administrative Functions

Several other administrative functions (which may be performed by other areas of the bank or by a third-party processor) support the overall operation, including report printing and distribution, purchasing, microfilm production, and supply room. Each must be given standards tailored to the particular service each function provides to the operation. All administrative functions are important to the performance of the whole, and each must include management's expectations for performance.

Summary

Several operating support functions have distinct roles. Typically they do not involve much, if any, direct customer contact; however, the way they are performed can greatly influence the level of customer satisfaction and affect customers' decisions about other banking services. At the same time, these operating support activities are important to the overall operation and, therefore, critical from management's perspective.

Review Questions

1. What are the two main accounting support activities?

2. What is the main activity of exception processing?

3. What is the distinction between incoming and outgoing chargebacks?

10
FRAUD

Bankcard fraud has been described as a two-man war game between the "good guys" and the "bad guys." The good guys are the issuing and acquiring institutions, the national associations, and law enforcement agencies. The bad guys are dishonest individuals or groups who are always on the lookout for ways to beat the system.

This chapter discusses bankcard fraud—what it is, how it occurs, and how the good guys combat it. Every organization that handles cardholder or merchant accounts has exposure to possible loss due to bankcard fraud.

After successfully completing this chapter, you will be able to

♦ understand the magnitude of bankcard fraud and its impact on bankcard profitability

♦ identify the most common sources of bankcard fraud losses

♦ list the main functions where exposure to fraud exists within the bank

♦ explain the importance of screening new merchants and monitoring merchant performance

♦ identify effective techniques for the detection, investigation, and successful prosecution of bankcard fraud

Magnitude of the Problem

The amount of dollars lost to fraud in the bankcard industry across the country has grown dramatically since the beginning of the 1980s. The American Bankers Association (ABA), MasterCard International, Visa International, and other interested parties in the credit card business have undertaken many efforts to help financial institutions find more effective ways to prevent bank credit card fraud. These parties work to stop fraud in instances where preventive measures fail. New technologies help in the effort and procedures are continuously being refined.

However, fraudulent transactions actually occur between people. Sometimes an honest person in the transaction, such as unsuspecting retail clerk or a bank teller, is fooled by a dishonest person. Other times, the transaction occurs between two dishonest people, such as a person working for a bank in collusion with a merchant. People are involved in every fraudulent bank credit card transaction.

A great deal of time and money are devoted to deterring criminals from altering or counterfeiting cards in an attempt to stop the dishonest user before he or she presents a card to a merchant or deposits a draft. Automation of merchant accounting and authorization systems has required major expenditures to help deter fraud. Yet, no foolproof method exists. As a result, these systems and procedures depend upon the people who use them to be effective.

Approximately 450,000 bankcards were reported lost or stolen in 1990. This is a large number in absolute terms, but it represents less than 0.5 percent of the cards in force. Of this number, less than 15 percent were actually used fraudulently.

In 1980, national bankcard fraud losses reported to the MasterCard and Visa associations amounted to a few million dollars. By 1991, the reported fraud losses amounted to about $198 million nationally. These losses come directly out of bank earnings, a concern that reaches the highest level of bank management. Approximately 60 percent of fraud losses occur before bank personnel know that the customer's card is missing. In fact, banks post approximately 40 percent of the fraudulent transactions to their customers' accounts before cardholders report their cards missing. Thus, banks must work harder to reduce the millions of dollars lost to fraud each year.

In terms of lost profits, fraud losses over the past several years have amounted to an average of 30 basis points. At first glance, this does not appear significant. However, as we have seen in the cardholder and merchant profitability models in chapter 3, this represents about 10 percent of bankcard profitability.

Major Sources of Fraud Loss

Criminals employ several methods to obtain and use cards fraudulently. These methods include the use of

♦ fraudulent applications that result in accounts being set up and cards issued to criminals

♦ lost and stolen cards used for unauthorized purchases

♦ counterfeit cards

♦ lost and stolen cards altered for fraudulent use

♦ collusive merchants engaged in fraudulent transactions using counterfeit and altered cards, white plastic fraud, and laundered drafts

♦ employee fraud in which employees steal cards from inside the card center (bank or third-party processor), give out valid account numbers to criminals, or set up bogus cardholder or merchant accounts

laundered drafts
Fraudulent transactions that are mixed with legitimate transactions for bank deposit in an effort to hide the fraudulent activity.

The Fraudulent Application

This method entails submitting bogus applications to the bank with the intention of establishing a credit line and obtaining cards for fraudulent use. With the volume of applications processed every month by banks across the country, early detection of a fraudulent application is key to the prevention of fraud.

Depending upon the sophistication of the perpetrator, detection of fraudulent applications can be extremely difficult. A manual application-processing environment frequently makes the effort to validate application authenticity more cumbersome. But every effort needs to be made to stop the fraudulent application before the account is set up.

Prevention

Many banks use fraudulent application files in either a manual or an automated environment to help detect the fictitious application. The fraudulent application file, unfortunately, sometimes allows the first fraudulent application to get through, but subsequent attempts that use some of the same information can then be detected.

Dishonest applicants frequently follow the same patterns in submitting applications. For example, the same address will be used for several applications, the same telephone number, or the same cardholder name (or the first or last name combined in different ways). By establishing a file that records the application information for each known fraudulent application, subsequent attempts to submit such applications are more likely to be detected.

All applications must be checked against the fraudulent application file to see whether any of the information on the application matches. When a match occurs, the security department is called to investigate the application and the applicant.

Lost or Stolen Cards

There are several concerns associated with a lost or stolen bankcard, and circumstances vary with individual instances and carry different kinds of risk. For example, if a cardholder has been robbed or burglarized, there is a greater likelihood that the card loss will be reported right away. Then the bank can get the card on the hot card file and reduce the possibility of fraudulent use.

hot card
An account on which excessive purchasing is taking place, which may indicate the card is being used for unauthorized purchases.

But some types of theft can remain undetected for weeks: the cardholder who leaves the card in the glove compartment of his or her car and doesn't notice it's missing for some time, or the card that is left on the counter at a merchant's store after a purchase and isn't missed for several days. The longer the time, of course, the greater the possibility of more losses to the bank. An unauthorized user of the card could have several days or weeks to use the card before it is listed on hot card files or the restricted card list.

Prevention

Prevention is difficult in the case of a lost card. The cardholder must take the initiative to report the loss to the bank. Too many times, steps taken by the financial industry have been met with apathy, if not resistance, on the part of customers who feel that the cost or burden of fraud does not fall on their shoulders. Some customers, for example, postpone opening their credit card bills because they know they do not have to pay them immediately. Thus, they neglect to examine their credit card statements in time to notice irregular charges. Although most cardholders are conscientious about keeping track of their cards and report a loss promptly when they become aware of it, some are not. The risk increases as more time elapses.

The limitation of cardholder liability to a maximum of $50 has also removed some of the urgency of reporting a lost card. Nonetheless, the best preventive measure is to educate cardholders to know where their cards are at all times. Consumers are the first line of defense against bankcard fraud. Periodic statement inserts reminding cardholders that everybody suffers when a card is lost and misused can be helpful. Printed messages on the monthly cardholder statement can also encourage consumers to be conscientious. Some banks promote the services of companies that report lost cards for the customer.

Counterfeit Cards

Counterfeit cards are facsimile cards embossed with false or fraudulently obtained account numbers and names. Detection of counterfeit cards is difficult at best. The variety of bank designs on MasterCard and Visa cards makes counterfeits tricky to spot. Merchant sales personnel are not experts in the detection of counterfeit cards. Also, as businesses try to speed customers' transactions through the checkout, the possibility that a counterfeit card will get through increases.

In some instances, collusive merchants knowingly accept counterfeit cards but claim they are innocent victims. If a merchant frequently accepts fake cards, it is reasonable to investigate whether the merchant is indeed a victim or a knowing party to the fraud.

Prevention

Other than visual detection, which is difficult, the best preventive measure uses authorization terminals in merchant locations and places the merchants on a zero floor limit. While it is unreasonable to expect 100 percent of the merchants to be on terminal authorizations, merchants with greater transaction volumes or greater exposure to counterfeit transactions should be targeted for authorization terminals.

Altered Cards

Like the counterfeit card, the altered card is not under the direct influence of the bank's control. Altered cards have been stolen from legitimate cardholders or from the bank. Authentic bankcards can be altered by changing account numbers, names, expiration dates, or all three. A wrongdoer may use several methods to make a card appear to be authentic.

Alteration without Merchant Collusion

In most cases, criminals alter cards and present them to legitimate merchants for purchases. The common techniques for altering cards include

◊ shaving off, rearranging, and gluing numbers or letters back onto a card
◊ heating a card to flatten the numbers or letters and then reembossing with different numbers or letters
◊ flattening numbers or letters with a machine
◊ punching holes around and then rearranging the numbers
◊ altering the signature strip

Alteration with Merchant Collusion

In some cases, collusive merchants accept altered cards. Merchant collusion is practiced when a dishonest store owner or other retailer knowingly accepts an altered or otherwise fraudulently obtained card. Merchants primarily use the following methods to participate in altered-card abuse:

◊ accepting cards known to be stolen and altered but not yet printed on the merchant warning bulletin or listed on a hot card file
◊ accepting cards that have been cut in half and rejoined with other card halves to form new account numbers
◊ creating multiple drafts for each altered card

Prevention

The following measures can be employed to prevent merchants from accepting altered cards:

♦ Place authorization terminals and authenticators in merchant locations to increase detection of altered cards.

♦ Meet with sales personnel of merchant outlets where the risk of occurrence is high. Bankcard security staff can show samples of altered cards to make sales personnel aware of what altered cards can look like and to explain the common characteristics of altered cards.

♦ Educate customers to guard their cards carefully and to report missing cards immediately.

♦ Closely monitor merchants that are repeatedly involved in altered-card transactions. A more aggressive chargeback policy or reward system may raise merchants' level of concern to prevent these transactions.

♦ Instruct merchants in the procedures for reporting suspicious cardholders through the authorization process.

Fraud Deterrents

Security experts in the bankcard industry continue to evaluate existing and potential security features for cards to improve the deterrence of credit card fraud. Banks and card manufacturers may use any combination of the following security features in their cards to help deter fraudulent use:

♦ Embedded ink, visible only under ultraviolet light, helps detect counterfeit cards—for use selectively with merchants that have a high exposure to fraud.

♦ Fineline printing, like that which appears on currency, makes counterfeiting more difficult because of precision printing.

♦ Microprinting of bank identification codes on the cards makes counterfeiting more difficult.

♦ Trademarks visible only under ultraviolet light reveal counterfeit cards—for use selectively in merchants with high exposure to fraud.

♦ Embossed security symbols facilitate the identification of the bank issuing the card (often another bank's customer account number will appear on an altered card) and make the tracking of drafts possible through easier identification.

♦ A printed bank identification number (BIN) above the embossed number makes alteration of the BIN on the card readily identifiable.

♦ Signature panels on the front of cards makes an erased signature more obvious to sales personnel. It is not uncommon for sales personnel to fail to check the signature on the back of the card with the signature on the sales draft even though merchant agreements require them to do so.

♦ Matched indelible account numbers on the back and front of cards makes counterfeiting more difficult and alterations easier to detect.

The Bad Guys

As mentioned earlier, fraudulent transactions occur between people. Let's look at who these individuals or groups are and how they operate.

The Collusive Merchant

The collusive merchant, like the dishonest employee, can cause substantial losses to the bank. This kind of merchant deliberately uses the trust implicit in the agreement with the bank to defraud. Methods by which merchants effect fraudulent transactions include

◊ deliberately accepting counterfeit cards
◊ deliberately accepting altered cards
◊ engaging in white plastic fraud—accepting drafts created by embossing valid cardholder account numbers on blank pieces of plain white plastic cards and presenting the drafts as legitimate transactions for deposit with the merchant bank
◊ engaging in the laundering of drafts—in this case, mingling fraudulent drafts generated by another merchant with legitimate drafts and depositing them with the merchant bank
◊ using valid account numbers for bogus mail order and telephone order transactions in which drafts are created for deposit with the merchant bank (telemarketing schemes have become widespread in recent years for this kind of fraud)

Collusive merchants pose an extremely high risk of loss to the bank. In some cases, a single collusive merchant has cost a bank several million dollars in just a short time. The bank unwittingly accepts the merchant's deposits, which are withdrawn before the bank is notified that the drafts are invalid. Hundreds of thousands of dollars can change hands in just a few days.

Employee Fraud

A dishonest employee, of course, carries great risk to the bank. Some of the ways a dishonest employee can defraud the bank include

◊ establishing fictitious cardholder accounts and issuing cards for fraudulent use
◊ stealing cards (blank stock or embossed cards) for fraudulent use
◊ stealing valid account information for embossing or encoding on counterfeit, altered, or white plastic cards
◊ establishing fictitious merchant accounts for handling collusive merchant deposits

When you think about the functions at your bank that frequently have access to live cardholder account information, you realize that it is virtually every area of the bankcard organization. Consider how this information could be abused and what preventive measures would be appropriate to limit this exposure to loss.

Security procedures are needed to ensure that all records containing customer information are accounted for when in use, are stored safely when not in use, and are accessible only to authorized personnel.

Fraud by Mail and Telephone

The criminal in this instance thrives on a growing number of mail order houses and telephone sales operations across the country that serve the public. Armed with valid account information, criminals easily order merchandise. If the transaction is accepted, with or without authorization, the merchandise is shipped. Often, the criminal takes little risk. Most mail and telephone order firms have agreed to accept chargebacks for unauthorized charges as a part of their cost of doing business.

Fraudulent Telemarketing

Telemarketing is another scheme that capitalizes on the absence of a cardholder signature or impression of credit card information (not required when the transaction is carried out by mail or telephone). In telemarketing schemes, the criminal telephones the cardholder and persuades the cardholder—often with too good to be true offers—to divulge account numbers and expiration dates. The scam proceeds with the completion and deposit of bogus sales drafts. Criminals quickly withdraw the funds and leave town before the fraud is detected.

Investigation of Fraud

When fraud is suspected, the first step is to undertake an investigation. The bankcard industry works closely with law enforcement agencies, and this cooperation has been the key to success during many criminal investigations and prosecutions.

The majority of industry security personnel belong to the International Association of Credit Card Investigators (IACCI). This trade organization has established a close, working relationship with many law enforcement agencies through its regional chapters worldwide.

Detection and Reporting

The bankcard industry continues to establish and refine early warning systems that increase the chance of apprehending the wrongdoer at the merchant point of sale. Familiarity with these systems and procedures helps the bank to detect potential fraud quickly .

Verification of Receipt of the Card

If a newly issued card is stolen before it reaches the legitimate cardholder, some time may expire before the bank receives unauthorized charges, and several weeks may pass before the true cardholder receives the monthly statement and reports the fraudulent transactions. Long delays make apprehending the criminal more difficult because the perpetrator may have left the area. And although witnesses may have seen the transaction occur, positive identification of the criminal is unlikely.

Many banks send the cardholder a second mailer requesting a call if the card has not arrived. This procedure greatly accelerates the reporting of stolen cards and lowers the bank's exposure to loss.

Lost and Stolen Cards

Several valuable resources help law enforcement officials apprehend criminals involved in the fraudulent use of cards, including

◊ reliable informants
◊ identification of other criminals who may have ties to organizations involved in the illegal use of lost or stolen cards
◊ identification and questioning of suspected clerks and tellers (often identified through initials and handwriting on sales drafts)
◊ photos obtained from surveillance cameras
◊ fingerprint analysis

Liaison with Law Enforcement Agencies

An institution's staff can assist the law enforcement agency during its criminal investigation by providing the following documentation:

♦ original sales drafts or true copies of drafts retained by the merchant and the bank

♦ authorization logs that can pinpoint the date and time of a fraudulent transaction

♦ an affidavit of forgery, usually obtained by deposition from the legitimate cardholder

♦ verification files and interchange records provided by the card-issuing banks

♦ reports developed by bankcard investigators

Prosecution of Fraud

Until the mid-1980s, bankcard fraud was not considered a significant law enforcement problem. Prosecutors rarely proceeded with felony convictions against those arrested with lost, counterfeit, or stolen cards in their possession. But the severity of the problem reached the point at which it demanded aggressive prosecution. What makes this specialized form of criminal activity so dangerous to society is that it provides the platform for more traditional forms of underworld activity: theft, embezzlement, possession and sale of stolen property, computer crimes, wire and mail fraud, and a host of other crimes.

Reacting to the growing threat of bankcard fraud, Congress broadened the scope of the Credit Card Fraud Act of 1984 to include fraud and related activities in connection with all payment devices. Now the counterfeiting and altering of credit cards is a federal offense. Several states have followed suit with special legislation aimed at the same broad interpretation of the law. Let's consider some of the tools and techniques used in the successful prosecution of bankcard fraud.

Arrests

In some cases, bankcard security representatives and law enforcement personnel may need a warrant for the arrest of a suspect. As is the precedent in other "economic" crime cases, proof of property ownership is generally required in an application for a warrant. For this purpose, the bank is considered an owner of a bankcard or credit card.

After the arrest of a suspect in possession of a card alleged to be fraudulent, the arresting officer may have difficulty contacting the legitimate cardholder for an affidavit before the suspect's arraignment. Bankcard security personnel may be able to supply the necessary information. An affidavit from the bank is usually sufficient to draft a complaint. If the arresting officer cannot reach a bank representative, the complaint may go forward on the officer's information and allegation that the card is lost, stolen, or counterfeit.

Criminal Complaint

Prosecutors may press several charges in a criminal complaint. Among these are larceny, criminal possession of a forged instrument, and possibly wire and mail fraud. A number of these criminal acts are covered by the federal Racketeer Influenced Corrupt Organization (RICO) Act and by related state statutes.

Regardless of the charges included in the complaint, the prosecuting attorney may need to work with a bankcard security investigator to get the documents necessary to present the case at arraignment or at a preliminary hearing required by a state or federal jurisdiction. Bankcard security personnel must ensure that customer information provided to the courts complies with the provisions of the Right to Financial Privacy Act. If the charge involves fraudulent use of the card, a witness—possibly the person to whom the card was presented for payment—may be required. Usually, the arresting officer can provide a statement from the witness.

Grand Jury Indictments

In most jurisdictions, the next stage of prosecuting a felony complaint involves a grand jury. However, in some jurisdictions the grand jury hears the complaint before arrests take place.

In more complicated cases, such as those involving a sting operation, it often helps to have a representative from the bankcard industry make a presentation to the grand jury about the fraudulent transaction process. Counterfeit cases may also benefit from the testimony of an expert witness.

Obtaining Necessary Documentation

A subpoena may be required to obtain documentation of the alleged fraud. Documents subpoenaed for grand jury presentation will be admitted under the rules of evidence related to business records. In some cases, original sales drafts are kept only for a limited time before they are recorded on microfilm, usually from 90 to 120 days. Although microfilmed records may satisfy the "best evidence" rules, in cases where handwriting is critical to obtaining a conviction, handwriting experts are generally reluctant to give an opinion on such evidence. This can present a problem if no other credible evidence exists to warrant prosecution.

Bankcard Industry Support

During the course of a grand jury inquiry, bankcard industry representatives can and should be contacted for information and support in obtaining evidence and documentation—and for any other reasonable assistance during presentation of the case. Among helpful items available from the industry are sample indictments for use in the prosecution of all types of card fraud cases. In addition, industry representatives can provide visual aids, which are an invaluable tool during fraud trials that involve counterfeiting or the alteration of cards. Visual aids and slides may also be used to acquaint jurors with the card transaction process. Familiarity with these procedures helps reveal how criminals take advantage of the system. This is particularly useful in the new and growing area of telephone order and telemarketing fraud.

Preparation and Conduct of a Trial

The bankcard industry can also help simplify the prosecutor's preparation for a trial. Security personnel often have a legal or law enforcement background and are a strong resource for overburdened prosecutors. An able prosecutor who proves to be well prepared for a trial can effect a conviction by a plea of guilty without a formal trial.

Summary

As the profitability of bankcard programs has increased, so too have the incidences of fraud. The national associations, the American Bankers Association, and federal law enforcement agencies are taking a much stronger stance in both prevention and prosecution of credit card abuses. The areas open to abuse are many: the forgetful cardholder who can't remember when he last used it and now reports it missing, the counterfeiter who alters lost or stolen cards, the collusive merchant who deliberately defrauds the bank. These are just some instances of how pervasive the problem is.

Review Questions

1. Approximately what percentage of Visa and MasterCard cardholders report their cards missing each year? Is it true that about 60 percent of fraud losses occur before the customer's card is reported missing?

2. List three major sources of cardholder fraud.

3. What tools and techniques can be used to help merchants detect fraudulent transactions?

4. What is the impact of fraud on the profitability of issuers and acquirers?

11

LEGAL AND REGULATORY ISSUES

The field of consumer credit is heavily regulated, both by agencies of the federal government and the individual states themselves. This chapter will familiarize you with the major laws that affect the bankcard industry and the legal requirements for compliance. Bankcard staff should be familiar with the law and know when to seek assistance from more senior managers, an attorney, a correspondent bank, the state bankers association, the American Bankers Association, or perhaps even the relevant regulatory agency. Failure to comply with what are sometimes difficult and technical legal requirements can result in significant civil monetary penalties.

Bankcard management should also be aware of pending legislation and regulations, so that it can lobby for needed changes. An informed management can articulate to the public directly and through the media how proposed government actions will affect bank services, including the cost to the banking industry, merchants, and cardholders.

After successfully completing this chapter, you will be able to

♦ identify the institutional sources of law and understand the functions of each

♦ discuss the effect of usury and rate limits on banks

♦ list and describe the major federal laws affecting bankcard operations

♦ understand the important elements of a comprehensive compliance program

Regulatory Agencies

Many different types of financial entities are issuers of bankcards. For example, bankcard issuers can be national banks, state-chartered banks, thrifts, or finance company lenders. Different government agencies regulate each type of financial institution. Congress has appointed the Federal Reserve Board as the regulator that enforces the provisions of most laws affecting consumer credit.

The Federal Reserve Board has issued regulations with respect to many statutes affecting bankcards, including Regulation Z, which implements the Truth in Lending Act; Regulation E, which implements the Electronic Fund Transfer Act; and Regulation B, which implements the Equal Credit Opportunity Act. The Federal Reserve Board is authorized to issue formal interpretations of those regulations and has authorized its staff to issue staff opinion letters. In general, creditors that follow the opinions of the Federal Reserve Board are generally immune from liability. In recent years, the Federal Reserve Board has significantly restricted the issuance of opinion letters. Instead, it has chosen to explain and interpret the meaning of some of the regulations in the form of a commentary to the regulation. The commentary to Regulation Z tracks the regulation in a section-by-section format, while the commentary to Regulation E adopts a less formal, and less comprehensive, question-and-answer format. In each case, however, the agency has provided a convenient, well-integrated, easy-to-use reference to supplement the regulation. These commentaries are updated periodically by the Federal Reserve Board to clarify points on problems that have arisen and add new topics in response to inquiries from the public or members of the industry.

In addition, the Federal Reserve Board has special duties to enforce the federal consumer credit laws in those banks over which it has general examination authority—that is, state-chartered banks that are members of the Federal Reserve System. It is joined in enforcement responsibilities by the Federal Deposit Insurance Corporation (FDIC) with respect to state-chartered banks that are not members of the Federal Reserve System and whose deposits are insured by the FDIC, and the Comptroller of the Currency (OCC) with respect to national banks. These three bank supervisory agencies have the dual role of bank examiner and enforcement agency.

The Federal Trade Commission (FTC) enforces the consumer credit statutes described in this chapter for all nonbank lenders, such as retailers and finance companies. The Federal Trade Commission has engaged in a number of rule-making efforts, including several that affect creditors. These include, for example, eliminating the holder-in-due-course protection for creditors, the debt collection practices rule, the used-car sales rule, and the credit practices rule. While the FTC has no direct regulatory or enforcement power over banks,

Congress has required the Federal Reserve Board to impose a substantially similar rule with respect to banks once such a rule is imposed by the Federal Trade Commission.

EXHIBIT 11.1 Federal Regulators of the Financial Services Industry

Primary Regulators	*Institutions[1] Supervised*
Comptroller of the Currency	National banks
Federal Reserve Board (FRB)	State-chartered banks (FRB members)
Federal Deposit Insurance Corporation	State-chartered banks (not members of FRB)
Office of Thrift Supervision[2]	Savings associations
National Credit Union Administration	Federal credit unions
Federal Trade Commission	Consumer finance companies, mortgage bankers, and certain other creditors

1. Institutions listed, except finance companies, are assumed to be insured.
2. Formerly the Federal Home Loan Bank Board. The Federal Home Loan Bank System still exists and has members ranging from insurance providers to commercial banks. Each of these members is supervised by one of the federal agencies listed.

State Governments

Prior to 1968, consumer credit was regulated solely by state laws. Despite the passage of broad federal laws in the area of consumer credit in the late 1960s and 1970s, Congress has not chosen to preempt all state laws in the consumer credit area. As a result, state consumer credit laws cannot be ignored, unless there is an express federal preemption. State legislatures have continued to demonstrate a strong interest in consumer credit and all states have laws that directly affect bankcards. Given the interstate nature of many credit and debit card programs and the comprehensive and highly technical nature of the federal consumer laws, federal preemption of state consumer finance laws would seem logical and desirable. So far, however, except with respect to interest rates, Congress has not chosen to preempt state laws. In fact, as the federal statutes are now drafted, state laws affecting consumer credit are not preempted unless they are inconsistent with a federal law dealing with the same issue, and then only to the extent of the inconsistency. Furthermore, some of the statutes affecting bankcards provide that even an inconsistent state law remains in effect if it provides greater protection to the consumer.

Usury and Interest Rate Limits

Usury laws control the maximum interest rates lenders can charge on loans. This issue has always been extremely important for bankcard issuers. Historically, interest rate limits have been set by individual states, resulting in no consistent pattern of interest rate limits. They range from states whose laws impose no interest rate limitations on banks, to states with laws that cap rates at levels so low they make it difficult for a bankcard program to be profitable.

Nonbank lenders look solely to state law to determine the interest rate that applies to them. National banks, state-chartered banks, savings and loans, and federally insured credit unions are authorized to charge a rate of interest pegged to the discount rate on commercial paper, or the interest rate allowed by the laws of the state in which the financial institution is located.

With the advent of interstate lending, the location of a bank has taken on increased significance, as has the question of which state law governs when the bank and its borrowers are located in different states. This issue was the basis for a 1978 U.S. Supreme Court decision, which interpreted the National Bank Act to permit a national bank to charge all of its borrowers the interest rate the bank is authorized to charge in its home state, even if the law of the borrower's state imposes a lower limit. This is known as a right to "export" the interest rate.

To provide equal treatment to FDIC-insured state banks, insured savings and loans, and insured credit unions, the 1980 Depository Institutions Deregulation and Monetary Control Act granted those institutions similar interest rate authority as the National Bank Act gives to national banks. Since interest rates are primarily dictated by the law of the bank's home state, banks have shopped for desirable places in which to locate based on the state's interest rate policy. For example, many bank holding companies set up banks in South Dakota and Delaware, because those states were early leaders in the efforts to deregulate consumer credit by removing usury ceilings.

The topic of interest rate exportation is not complete without mentioning the controversy over what fees comprise a bank's interest rate for purposes of exportation—the finance charge alone or other account fees, such as annual membership fees and late fees. The law on this issue is somewhat unclear, and a few states have sued out-of-state lenders to attempt to force them to apply the law of the cardholder's state rather than the bank's state.

Federal Statutes

Truth in Lending Act

The oldest and best-known consumer credit statute is the Truth in Lending Act (TILA), which was enacted in 1968 as part of the Consumer Credit Protection Act. The Federal Reserve Board's Regulation Z implements the provisions of TILA. It applies only to consumer credit, which is credit extended to an individual for personal, family, or household purposes. TILA governs open-end revolving credit, such as credit card loans, as well as closed-end credit, such as mortgage loans. We will only discuss the aspects of TILA that apply to bankcards.

TILA's purpose is to require lenders to provide consumers with meaningful disclosure of important credit terms so that they will be able to shop for credit and have enough information to make an informed decision. TILA also establishes procedures for the timely resolution of credit billing disputes and prohibits the issuance of unsolicited credit cards.

TILA fundamentally is a disclosure statute—it does not regulate the terms under which a bank offers credit, it only requires that certain key loan terms be disclosed in specific ways and at prescribed times. Based on a 1990 amendment to Regulation Z to implement the Fair Credit and Charge Card Disclosure Act, creditors must now disclose specified credit terms in a segregated box on or with each credit application or solicitation for consumer credit. A second disclosure, traditionally known as the "initial disclosure statement," must be made before the first transaction can be made on a bankcard account. This disclosure generally is combined with the credit agreement that is sent to the consumer. In addition, the bank must provide the consumer with monthly billing statements that summarize account activity and provide other basic information, and a monthly or annual statement of the consumer's rights with respect to billing errors.

Other sections of TILA prohibit banks from distributing credit cards on an unsolicited basis. As a result, banks may send out credit cards only in response to a request from a consumer or when an existing account is renewed. TILA also limits a cardholder's liability for unauthorized use of a credit card, which currently stands at $50, and regulates the advertising of credit terms.

Due to the numerous requirements imposed on creditors and the complexity of the TILA provisions, all creditors should have comprehensive compliance procedures that are prepared and reviewed by legal counsel. As a strict liability statute, a creditor is liable for any violation of TILA, and the penalties for even occasional or technical failures to comply can be high. Consumers can recover any actual damages they suffer plus punitive damages of up to

$1,000 per occurrence. In a class action brought on behalf of many consumers, the group can recover up to $500,000 or 1 percent of the creditor's net worth, whichever is less. In addition, if the consumer prevails, the court could require the bank to pay the consumer's attorneys' fees and court costs.

Fair Credit Billing Act

The Fair Credit Billing Act (FCBA), one section of TILA, requires creditors to investigate and resolve written inquiries from consumers concerning alleged billing errors on their accounts. The definition of a billing error in the FCBA is quite broad. It includes not only computational errors and misposted items but also inadequate transaction descriptions and the failure to mail a periodic statement to the cardholder's correct address.

The FCBA requires a consumer to notify the creditor of the alleged error within 60 days of the mailing of the billing statement. It then requires the creditor to investigate the inquiries promptly, resolve the dispute, and notify the consumer of the action the creditor will take within two complete billing cycles (up to 90 days). While the dispute is pending, the creditor must refrain from trying to collect the amount in question and is prohibited from reporting the cardholder's account as delinquent to a credit bureau.

Under certain circumstances, the FCBA also allows cardholders to withhold payment of merchandise or services if the consumer has a dispute with the merchant regarding the goods or services. The FCBA allows a cardholder to withhold payment for any damaged or inferior goods or services purchased with a credit card if

    ◊   the purchase price exceeds $50
    ◊   the transaction occurred in the cardholder's home state or within 100 miles of the cardholder's residence
    ◊   the cardholder has made a good faith effort to resolve the dispute with the merchant

This provision puts banks in the position of having to mediate disputes between cardholders and merchants concerning the quality of goods or services charged to a bankcard account.

Equal Credit Opportunity Act

The Equal Credit Opportunity Act (ECOA) is another section of the Consumer Credit Protection Act. ECOA prohibits discrimination in any aspect of a credit transaction based on race, color, religion, national origin, sex, marital status, age, or the fact that the applicant receives public assistance benefits or has exercised his or her rights under another section of the Consumer Credit Protection Act.

The nature of ECOA is more like a civil rights statute that is designed to prevent discrimination in the credit-granting process. Of course, making credit decisions is based on differentiating between customers who are expected to have a high potential for repaying a loan versus those for which poorer performance may be presumed. ECOA and the Federal Reserve Board's Regulation B, which implements it, do not forbid creditors from making judgments about which applicants for credit are deemed to be creditworthy. The law does, however, prohibit creditors from using any of the stated criteria (such as sex, age, race) as the basis for granting or denying credit.

Under certain circumstances, ECOA would also prohibit a creditor from using criteria that might be neutral and nondiscriminatory on its face, but would have the effect of discriminating on the basis of one or more of the prohibited classifications. For example, a bank's decision to deny credit to any person living in a particular geographical area (or having a particular ZIP code) would appear neutral on the surface but might have the effect of discriminating based on race or national origin. The application of the effects test in a credit context has not yet been tested in the courts, but it is likely that a well-constructed system for granting credit—where the credit-granting characteristics can be statistically shown to relate to creditworthiness—will be defensible against attack. Perhaps the most scientific basis for constructing such a system is the adoption of an empirically derived credit-scoring system.

ECOA applies to business and commercial credit and not just consumer credit, although some of the technical requirements are slightly different. In 1988, ECOA was amended by virtue of the Women's Business Ownership Act to remedy the continuing discrimination against women and businesses owned by women in the granting of credit. The amendments to Regulation B, which became effective in early 1990, require lenders to make notifications of the reasons for loan declines and disclosures to all applicants for business loans.

Regulation B specifies the type and form of information that can be collected from an applicant for credit and how that information can be used. For example, creditors cannot ask an applicant's marital status and cannot close an account because a borrower becomes divorced or widowed. Consequently, a bank must carefully design its credit application forms and thoroughly train its loan officers. ECOA also requires lenders to notify a consumer whose request for credit is denied. The notice must be sent within a prescribed time and must list the specific reasons for the denial of credit or offer to supply the reasons upon request. The regulations also dictate under what circumstances a creditor can require that another party act as a cosigner on a loan, if the lender believes that the applicant does not have the appropriate credit standing to qualify individually.

ECOA's prohibition against discrimination in the granting of credit covers every aspect of a creditor's activities from advertising of the availability of credit to debt collection. While ECOA is in some respects easier to comply with than the highly technical TILA, its subjective nature presents different compliance problems; disclosure rules are sometimes easier to comply with than a statute with more conceptual requirements. Banks should evaluate their credit programs to ensure that they are nondiscriminatory. An internal audit and compliance function can review forms and procedures to protect the bank and bankcard operation against exposure to unlawful practices.

Fair Credit Reporting Act

The federal Fair Credit Reporting Act (FCRA) is yet another section of the Consumer Credit Protection Act. Its principal objective is to regulate the purposes for which credit bureaus are permitted give out information about consumers from their files.

The FCRA requires credit bureaus to ensure that the information they give out is accurate and to delete obsolete information after a prescribed time. The FCRA does not apply to applications for business credit or to data about businesses that are maintained at credit bureaus. No regulations have been issued in connection with the FCRA. Unlike TILA and ECOA, Congress has authorized the Federal Trade Commission to issue interpretations of the FCRA.

The FCRA imposes civil penalties on anyone who willfully or negligently fails to comply with the FCRA, as well as allows recovery by the consumer for actual damages incurred.

One of the most important aspects of the FCRA is that a credit bureau must verify that a person requesting credit information on a consumer has a permissible purpose for receiving that information. When a consumer applies for credit, a permissible purpose exists. If the consumer does not initiate the request for credit, a user will only have a permissible purpose if the information will be used in connection with a credit transaction involving the consumer (such as for an extension of credit or a review of, or collection efforts on, an existing credit account), or if the user has a legitimate business need for the information in connection with a business transaction involving the consumer.

A highly controversial aspect of the FCRA is whether the practice of prescreening by creditors is permitted by the FCRA. Prescreening is the procedure whereby a creditor asks a credit bureau to review a list of consumers to be solicited for credit to determine in advance whether the consumers listed meet the credit standards of the creditor. Rejected names are deleted, and the creditor then solicits only the remaining names on the prescreened list. For many years, the FTC has held that prescreening is permissible under the FCRA only if a firm offer of credit is made to every person who passes the prescreening. The FTC reaffirmed this interpretation in its 1990 commentary to the FCRA.

Although the FCRA is aimed primarily at credit bureaus, it also imposes requirements on creditors, such as banks that use consumer credit information. For example, if a creditor denies an application for credit or increases the charge for credit based on information obtained from a credit bureau, the creditor must tell the denied applicant the name and address of the credit bureau that provided the information. Similarly, if credit is denied based on information provided to a creditor by any other person, such as another creditor or an employer, the creditor must advise the consumer in writing of his or her right to request the nature of that information.

The FCRA was enacted in 1970 and has not been amended. In recent congressional sessions, legislation has been introduced to correct perceived abuses in the use of credit information and to take into account current practices of credit bureaus and users of credit reports. Although no such legislation has yet passed, it is likely that the FCRA will be amended in light of the pressure of consumer groups and the adverse publicity concerning credit screening practices.

Fair Debt Collection Practices Act

The Fair Debt Collection Practices Act, which was adopted in 1977 as yet another section of the Consumer Credit Protection Act, is designed to eliminate abuse and deceptive debt collection practices. This act applies only to persons who regularly collect consumer debts owed to another person, and not to creditors collecting debts on their own behalf. Thus, a bank using its own name that engages in collecting debts owed to it is not subject to this act. However, the act does apply to banks that regularly collect debts for other unaffiliated financial institutions. For example, under a typical arrangement in which one bank helps another bank collect a defaulted debt of a consumer who has moved to another area, the collecting bank would be subject to this act. Also, a bank must comply with this act if it uses a name other than its own name in its collection efforts.

The act prohibits debt collectors from using abusive tactics or harassment in collecting debts. It also prohibits debt collectors from making false and misleading representations, such as creating the false impression that collection documents represent a formal legal process. The act limits the amount of contact that the debt collector can have with the debtor and with third parties, such as the debtor's employer. It requires the creditor to provide certain information concerning the debt to the consumer during the initial contact and to cease further communication with the consumer upon request.

The Federal Trade Commission has urged Congress to expand coverage of this act to include creditors that are collecting debts on their own behalf. In addition, a number of states already have laws prohibiting unfair debt collection practices that apply to banks and other creditors collecting debts in their own name. Therefore, banks must be particularly sensitive and attentive to the procedures used by bank personnel and by third parties used by the bank in collecting delinquent debts.

Electronic Fund Transfer Act

The most recent addition to the Consumer Credit Protection Act is the Electronic Fund Transfer Act (EFTA), passed by Congress in 1978. The Federal Reserve Board's Regulation E implements EFTA.

In many respects, EFTA is similar to provisions governing credit cards under the Truth in Lending Act. EFTA covers all transactions initiated electronically that authorize a debit or credit to an asset account, such as a checking account. It also prohibits unsolicited distribution of validated electronic funds transfer cards. EFTA requires financial institutions to make written disclosures to consumers

◊ when a consumer first contracts for an EFT service
◊ periodically (typically monthly), when electronic transactions have occurred
◊ before changing the terms of an account

EFTA also establishes procedures for resolving errors in electronic funds transfers. Under Regulation E, for example, the bank must provisionally recredit amounts that a customer claims are in error if 10 days have elapsed without the alleged error being resolved. EFTA also limits a consumer's liability for unauthorized electronic fund transfers, for the failure of the bank to complete transactions properly, and for system malfunctions. EFTA's civil liability provisions are similar to those found in the Truth in Lending Act.

Because the rules that apply to electronic funds transfers are somewhat different from those that apply to credit cards, there is sometimes confusion as to which law applies in connection with a card that has both debit and credit capabilities. Banks must evaluate carefully which law applies in a given transaction.

Bankruptcy Code

The federal laws governing bankruptcy procedures are found in the Bankruptcy Code. This code allows consumer debtors to bring before the courts two different types of cases, those under Chapter 7 and those under Chapter 13.

About three-fourths of all consumer cases are brought under Chapter 7. In a Chapter 7 case, sometimes called a straight bankruptcy case, a court-appointed trustee collects and liquidates the debtor's nonexempt property and distributes the proceeds to creditors. Creditors that have a security interest in the debtor's property generally receive either the security itself or its value. The debtor receives a discharge, which means the debtor is released from any legal obligations to pay debts.

The alternative, Chapter 13, is referred to as a plan for a debtor with regular income or as a wage earner plan. Under Chapter 13, the debtor generally retains all or almost all of his or her property and promises to pay all or a portion of his or her debts out of future income, in accordance with a court-approved plan.

A bankruptcy case begins with the filing of a bankruptcy petition by the consumer debtor. An important feature of both Chapter 7 and Chapter 13 cases is that, once the petition is filed, creditors are prevented by law from taking any action, formal or informal, to try to collect their debts. Therefore, every credit card issuer should establish internal procedures for communicating information about bankruptcy cases to avoid penalties for impermissible collection efforts. In addition, in Chapter 13 cases, the prohibition against collection efforts also applies to creditor action against a codebtor or coobligor, even if such a person is not in bankruptcy. Therefore, if both spouses are liable on a credit card account, the stay prevents a credit card issuer from taking action against the spouse who has not filed a bankruptcy petition to collect the debts of the filing spouse.

Under the Bankruptcy Code, the debtor may exempt whatever property is exemptible under the laws of the state in which he or she lives. In about 10 states, as an alternative to using state exemptions, the debtor may exempt property from a list of federal exemptions given in the Bankruptcy Code. State exemption laws vary widely, and we can identify only the most common types of exempt property. Most states and the federal exemptions exempt at least some of the following, often with a dollar limit:

◊   homestead
◊   furniture and household articles
◊   clothing
◊   tools of the trade and professional supplies

◊ automobiles

◊ health aids

◊ wages

◊ pensions and other similar sources of future support

◊ an omnibus dollar amount (sometimes as an alternative to the items already listed)

The main reason for a consumer to file a Chapter 7 petition is to receive a discharge. Since discharged debts are forgiven, creditors are enjoined from trying to collect them. Under the Bankruptcy Code, a debtor's promise to pay a discharged debt—called reaffirmation—is not valid unless approved by the court. Courts will rarely approve the reaffirmation of credit card debts.

discharge
Allowance under the Bankruptcy Code releasing the debtor from any legal obligations to repay debts.

Under the Bankruptcy Code, a number of debts are nondischargeable. If the court decides that a particular debt is nondischargeable, the debtor must pay that debt even though other debts are discharged. However, even a debt that is nondischargeable under Chapter 7 may be discharged if it is dealt with as part of a Chapter 13 plan.

For credit card issuers, the most important categories of nondischargeable debts are those incurred on the basis of fraud, false pretenses, or false financial statements. These situations, called loading up, include the following:

♦ The debtor incurs substantial debts shortly before filing the petition.

♦ The debts are for luxury items, not necessities.

♦ The debtor acquires the debts knowing that his or her overall financial situation is such that the debt can never be repaid.

♦ The debtor has already consulted legal counsel about a possible bankruptcy case.

♦ The debtor makes a large number of small purchases (under $50) to avoid credit checks.

♦ The debtor knowingly exceeds the credit limit on the account.

loading up
The practice of charging debt on open-end credit on the eve of bankruptcy.

The protection offered creditors by the Bankruptcy Amendments and the Federal Judgeship Act of 1984 is limited. Banks must be vigilant if they are to control losses due to bankruptcy at an acceptable rate. As part of a broad coalition of consumer creditors, banks have made a vigorous effort to amend the Bankruptcy Code and bring a greater sense of balance to the laws affecting consumer bankruptcies.

Financial Privacy

The federal Fair Credit Reporting Act is a manifestation of society's increasing concern about the collection and dissemination of personal information. The growth in technology and increased use of computer data systems have led to a concern for the preservation of the individual's right to privacy. While early legal discussions defined privacy as the right of a citizen to be left alone, the technological revolution has shifted the focus to the individual's right to control the circumstances under which, and to whom, personal information is disseminated.

One important aspect of the privacy issue concerns the government's access to personal records. The Privacy Act of 1974 regulates the collection and use of personal information by federal agencies and departments. This act also established the Privacy Protection Study Commission to study the computer data and information systems operating in both private and public sectors. The commission was to recommend to Congress what needed to be done to protect the privacy of individuals, while meeting the legitimate needs of government and society for information. The commission's recommendations were presented in its 1977 report, "Personal Privacy in an Information Society."

Another aspect of the privacy issue focuses on government access to records maintained by private, third-party recordkeepers. Information held by financial institutions may be particularly sensitive, since the records of an individual's financial transactions can be used to construct a detailed chronology of a person's activities and associations. This is particularly true of bankcard records, and the potential danger will increase as we continue to move into an electronic funds transfer environment.

Congress gave individuals protection from government access for the first time in 1976. In the Tax Reform Act of 1976, Congress singled out the agency most frequently apt to seek access to bank records, the Internal Revenue Service (IRS). The act provides certain protections to a person whose tax records are sought from a third-party recordkeeper, such as a bank, under an IRS summons. The person, when notified of the amounts in question, has the right to attempt to stay the bank's compliance with the summons and the right to intervene in any proceeding to enforce the summons. In addition, banks may receive reasonable reimbursement for their costs in compiling the data required by an IRS summons.

The Privacy Protection Study Commission, in its 1977 report, recommended that similar legislation cover all types of government requests for records. This recommendation ultimately led to the passage of the Right to Financial Privacy Act of 1978, which was enacted as Title XI of the Financial Institutions Regulatory and Interest Rate Control Act.

The 1978 Privacy Act forbids financial institutions from disclosing customers' records to government officials except in response to a written authorization from the customer or a written legal demand, such as an administrative summons or subpoena, a judicial subpoena, a search warrant, or a formal written administrative request. The Privacy Act also establishes certain procedures that financial institutions must follow in disclosing information in response to a written legal demand. These procedures include the following:

♦ The government agency or official demanding the records must first notify both the customer and the financial institution that a subpoena or other demand has been issued.

♦ A statutory waiting period is required to allow the individual whose records are sought to go to court to challenge the government's demand.

♦ During the statutory waiting period, the financial institution must assemble the requested documents but cannot release them to the government agency.

♦ If the statutory waiting period expires without challenge by the customer, the government agency or official must certify to the financial institution that the government has complied with the act's procedures. Only then can the financial institution disclose the information and, in certain cases, obtain reimbursement for its costs in compiling the requested information.

The Privacy Act of 1978 restricts financial institutions and other third-party recordkeepers with respect to their release of financial records to federal agencies only. Congress has considered several proposals that would impose additional controls on the collection, maintenance, and use of personal data by the private sector. Many of these legislative proposals grew out of other recommendations of the Privacy Protection Study Commission's report. This area is likely to receive considerable legislative attention in the years ahead.

General Compliance Guidelines

The legal and regulatory issues with which banks must deal are increasingly complex and constantly changing. All of the consumer laws discussed in this chapter were enacted by Congress between 1968 and 1978—a 10-year period of legislative flurry fueled by pressure from consumer groups—and most have been amended numerous times since. Neither congressional nor consumer interest in bank legislation appears to be waning, and banks can anticipate additional regulations from federal agencies. The coexistence of state and federal laws on the same subjects further increases the volume of law that must be monitored and mastered by creditors and their counsel. Rigorous enforcement of these laws by bank examiners and greater sophistication among consumers concerning their rights make a formal, comprehensive compliance program a necessity for any bank engaged in consumer lending. At a minimum, a bank's compliance program should include the following basic elements:

♦ A bank should designate one or more compliance officers, who have no direct responsibility for any particular lending program or product, to work closely with bank counsel to

  ◊ maintain a familiarity with consumer laws
  ◊ oversee and coordinate the bank's effort to comply with those laws
  ◊ review and respond to inquiries and complaints from consumers and regulatory authorities
  ◊ serve as a resource for answering questions concerning legal compliance from lending officers and other bank personnel

♦ A bank should prepare and frequently update written policies and procedures manuals for use by its lending officers and should supplement these with special bulletins when important legal developments warrant them. In particularly complex and/or far-reaching legal matters, bank counsel may be asked to brief officers.

♦ A bank should provide regular, in-depth training sessions on relevant consumer laws for all bank personnel involved in the bankcard operation, including operations, marketing, and all customer contact personnel.

♦ A bank should periodically review all agreements, disclosure statements, form letters, notices, and other forms used in consumer lending to ensure that bank documents comply with the law and are updated regularly.

♦ A bank should provide opportunities for frequent oversight of the internal compliance program by the auditing staff, bank counsel, and senior managers (who bear the responsibility for compliance and/or for the functions that must be in compliance with the various laws and regulations).

The cost and effort associated with compliance are high. The cost of violating laws, even inadvertently, can be higher still. The laws and the lawmaking process are far from perfect. Bank management should use the available mechanisms for changing the undesirable provisions of those laws, even while making an effort to comply with them. Many of the laws were created in response to consumer groups' demands. A management that is concerned and goes beyond minimal compliance may discover a marketing advantage and improved customer relationships that far outweigh the cost of compliance.

Summary

The period between 1968 and 1978 was highly active in terms of consumer legislation pertaining to financial institutions generally, with some specific focus on bankcards and the consumers who use them. The major federal laws that affect bankcard operations include the following:

- ◊ The Truth in Lending Act (Regulation Z)
- ◊ The Fair Credit Billing Act
- ◊ The Equal Credit Opportunity Act (Regulation B)
- ◊ The Fair Credit Reporting Act
- ◊ The Fair Debt Collection Practices Act
- ◊ The Electronic Fund Transfer Act (Regulation E)
- ◊ The Federal Bankruptcy Code

The requirements to comply with these laws and regulations are strict. Failure to comply can result in high penalties. Management should take appropriate steps to assign internal accountability for compliance and auditing of compliance. There is a significant cost to ensure compliance, but the costs of noncompliance can be greater.

Review Questions

1. What are the principal acts that affect the bankcard business and which agencies enforce their compliance?

2. You are the CEO of a state-chartered bank in a state with a low usury rate ceiling. If you want to make your bankcard portfolio more profitable by charging higher prices, what are your options?

3. Create a hypothetical compliance function in your bankcard operation and describe its activities.

12

THE OTHER CARDS

The personal bank credit card is by far the most widely used of the many plastic cards provided by financial institutions and is the main focus of this text. However, in addition to the bank credit card, several other types of cards have been developed. All offer convenience to consumers and merchants. Depending on the bank and the specific type of card, the responsibility for managing the functions associated with the different cards varies. For example, the administration of the check guarantee card may be assigned to the bankcard department or another division of the bank.

Chapter 1 briefly discussed the other types of cards in the marketplace. Now let's look at them in more detail.

After successfully completing this chapter, you will be able to

♦ explain the function of each of the other cards

♦ describe how a debit card differs from a bank credit card

♦ list the benefits of private label cards to merchants

♦ discuss the continued viability of these other cards

ATM Card

ATM is an acronym for automated teller machine. Most ATM cards enable the authorized holder to perform the following functions:

◊ withdraw cash from checking and savings accounts
◊ deposit to checking and savings accounts
◊ obtain a cash advance from a MasterCard or Visa account
◊ make a loan payment, such as to a bankcard, automobile loan, or real estate loan account
◊ get balance information on checking or savings accounts or the available credit on a bankcard or other credit account, such as a line of credit attached to the checking account
◊ transfer funds from one account to another, such as from a savings account to a checking account

The ATM card extends banking convenience to the customer. Because ATM machines are typically on the exterior of the bank, or in some cases, at a location away from the bank, they usually operate around the clock, seven days a week. Therefore, the customer can access accounts without having to go into the bank. Consequently, the ATM card is sometimes called an access card. This means, for example, that the customer can use the ATM to get money without searching for somewhere to cash a check. Deposits and loan payments can be made at night or on weekends, and the customer can get his or her account balance even when the bank is closed.

Consumers have responded favorably to the ATM card. Although use of the card is not universal, those who decide to use it do so actively. Once the customer is familiar with the machine, he or she frequently uses it in place of visiting the bank. Generally, ATM users tend to be younger and more mobile socially and geographically, while older customers seem to prefer going into the bank to conduct their business with an employee.

The popularity of ATM cards persuaded several banks to form competitive alliances to expand the geographic reach of the card service. Regional networks formed to offer ATM access outside the customer's immediate area of residence. Cirrus and Plus, two national ATM networks, followed with alliances of banks across the country to give ATM cardholders access to their accounts from many places throughout the United States. For example, a cardholder visiting California can access a checking or other account with a bank at home in New York. However, if the ATM card is a part of a proprietary system (like Cirrus), the cardholder must use the card at a bank that is a part of that network. Banks in nationwide systems provide cardholders with toll-free telephone numbers and printed directories of nearby ATM machines. These are especially helpful when cardholders are traveling.

During the early years of ATMs, the cost-avoidance benefits of the new card received a lot of publicity. Bank managers expected many customers to use the card and the bank could then reduce the number of tellers inside the bank. Others believed the same number of tellers could be maintained, while the customer base would grow because many customers would use the machines instead of coming into the bank to do business. However, these expectations went largely unmet. By 1990, there were about 140 million cardholders in the United States. About half of American households hold at least one ATM card, and more than half use their cards at least once a month.

Bank marketing focused attention upon the convenience of the card. The advertising theme, directed at people who were likely to use the ATM card, trumpeted extended banking hours. These efforts worked reasonably well, and banks began to add large numbers of ATMs to attract that segment of the market. Banks in branch banking states—states that permit banks to operate branch offices—promoted the geographic convenience offered through ATMs dispersed throughout wide areas. Banks in unit banking states—states that require banks to operate without branch offices—began to band together to share machines so that their customers could enjoy the convenience of expanded geographic access. Shared networks were promoted as a reason to "bank with us."

The operating systems supporting ATM programs are generally separate from those supporting MasterCard and Visa programs in banks. At the outset, the ATM card was more closely tied to a checking or savings account than to a credit card account. To some extent, the ATM operating systems and the credit card operating systems have moved closer together. But many banks keep them separate, both functionally and administratively. In the consumer's mind, however, the ATM card is largely viewed as an additional bankcard that offers another form of convenience.

Check Guarantee Card

Because the check guarantee card is frequently attached to a line of credit associated with a personal checking account, it is sometimes not considered a bankcard. However, the check guarantee card is plastic, issued by a bank, and presented to a merchant to validate a purchase. Therefore, consumers and merchants generally consider it another bankcard.

To the consumer, the check guarantee card offers another form of banking convenience. Checks can be cashed at merchants' stores more easily. In fact, check cashing in many areas of the country is virtually impossible without a check guarantee card or major credit card, such as MasterCard or Visa. These cards offer the merchant some assurance that the check being presented is valid. The customer must request a check guarantee card attached to a personal line of credit, because unsolicited issuance violates federal law. Certain disclosures required under Regulation Z must also accompany issuance.

The distinctive feature of a true check guarantee card is that it does guarantee the check. If a bad check is returned, the merchant may collect the amount of the check from the bank that issued the card. However, most check guarantee cards may only be used to guarantee first-party personal checks and not payroll checks or checks made payable to the person holding the card.

Major credit cards accepted in place of a check guarantee card are generally used as a means of identification. This offers the merchant a higher degree of confidence that the check presented by an individual will be honored.

Banks typically do not charge a fee for the card or check guarantee service. If credit is used, the customer pays the interest and fees associated with the personal line of credit. Thus, most consumers view the check guarantee card as a free service that makes check cashing easier.

From the merchant's point of view, the value of a true check guarantee card is that payment is assured. To effect the guarantee, the merchant must follow the terms printed on the back of the card or comply with the conditions of contract where appropriate. (A few banks have contractual agreements with merchants for the card.) In either case, the terms of guarantee are generally simple:

♦ The check must be a personal check presented by the check guarantee cardholder.

♦ The amount of the check cannot exceed a specified amount ($200, $500, and so on).

♦ The expiration date on the card must not have passed.

♦ The signature on the card must be reasonably similar to the one on the check.

If the specified procedures are followed, the merchant enjoys the benefit of guaranteed payment. However, some merchants require an additional form of personal identification, such as a major credit card or driver's license, as an added precaution in case the merchant must try to collect payment from a person who wrote a bad check.

From the bank's perspective, the check guarantee card extends convenience by facilitating check cashing. The card is another means of offering credit to consumers and helps consolidate the relationship with the bank. The check guarantee card also promotes the bank's identity in the marketplace when the card is presented to merchants. In the early 1970s, banks heavily promoted check guarantee cards in various parts of the country. Since then, the costs of administering these programs and indifference by many merchants toward check guarantee have caused the level of attention among banks to subside in favor of bank credit cards or debit cards.

Debit Card

The point-of-sale (POS) debit card is really a combination of a check guarantee card and an ATM card. If a checking account transaction can be performed at an ATM, why not have the same function performed at a merchant location (provided the merchant has the equipment necessary to accommodate the card)? The debit card can be used in place of a paper check and the transaction will be automatically guaranteed because funds transfer immediately from the purchaser's account to the seller's.

Debit is, of course, a financial term. Its use in connection with the card implies access to a deposit account, as opposed to the line of credit accessed by the bank credit card. Although we refer to the debit card as a generic bankcard, to be precise, some distinctions should be made:

♦ **Proprietary Debit Card.** This card identifies a specific bank or a group of banks in a regionally shared point-of-sale network. The Honor System in Florida, the New York Cash Exchange (NYCE), and the Interlink System in California are examples of regional networks. In virtually all cases, proprietary debit card transactions were initially handled outside the conventional national credit card networks. Recently, however, developments in systems and operating rules have allowed transactions to flow through the MasterCard and Visa networks. Also, the label of proprietary debit card has been relaxed and refers to bank ATM cards as well.

♦ **Point-of-Sale Card.** This refers to any card presented at the point of sale—the merchant's store or other location away from the bank. The POS system uses communication lines and is designed to authorize, record, and forward electronically each sale as it occurs.

Let's look more closely at how the debit card functions. As noted, the term debit refers to accessing a deposit account—typically a personal checking account, although the card can access a savings or a money market account. When used to make a purchase at a store, the debit card takes the place of a personal check. The record of the transaction appears on the customer's checking account statement. To validate the sale, the merchant follows authorization procedures much like those followed for credit card purchases.

In spite of immense attention paid to debit cards in the 1970s and early 1980s, widespread use in the marketplace is just beginning. The growth of proprietary debit cards is accelerating because many supermarkets and other high-volume check-cashing merchants are beginning to accept them. As mentioned earlier, proprietary debit cards are frequently issued by banks participating in regional point-of-sale networks. Transactions are handled outside the national networks, and the cost of interchange, as applied to bank credit card transactions, is avoided. Therefore, with greater merchant acceptance, more consumers seem willing to use the proprietary debit card.

One factor that limits the appeal of the debit card involves the consumer's choice to use either personal or bank funds to pay for a purchase. Not only does a debit card access personal funds, it also effects immediate transfer of funds from the account, and so the period of processing and collection—known as float—is eliminated. Nationally, about 30 percent of bank credit card users prefer to pay their balances in full each month. But because many banks offer a grace period with their credit cards, the holder may have more than 30 days to pay the bank for the purchase. Thus, consumers have favored the use of credit cards for payment.

All things being equal, merchants that accept bank credit cards should be willing to accept debit cards. A merchant that follows routine authorization procedures for the card validates the sale and guarantees its payment. In addition, the forms and procedures used for handling debit cards are similar to those used for credit cards. Therefore, merchants are already familiar with the routine. However, not all things are equal.

First, some merchants view the cost of interchange as more expensive than the cost of handling a personal check. Research studies have indicated that handling personal checks is more expensive than the cost of interchange, but this research typically comes from banks. Other studies indicate that the cost of interchange exceeds that of handling checks, and this research typically comes from merchants. Unfortunately for banks, the merchants appear to be winning the argument. The fees merchants pay for proprietary debit card transactions are generally less than the cost of interchange on bank credit card transactions. The merchant's view is fundamentally that of the pragmatic businessperson: given a choice between the cost of a personal check or a debit card, the merchant will take the less expensive transaction.

A second issue is that of paper transactions versus electronic transactions. Merchants generally view the latter as less expensive because checks have to be handled from the store to the bank. In addition, the merchant must pursue collection on returned checks. In the case of an electronic transaction, the paper stays on site (the customer gets a receipt and the merchant retains a copy of the transaction). Also, payment to the merchant is basically guaranteed and immediate. Use of the proprietary debit card in supermarkets and other high-volume businesses is almost exclusively electronic. Again, the merchant seeks a less expensive way to validate the sale and process the transaction.

A third issue, especially for high-volume merchants, involves the speed of checkout. Electronic debit card transactions are usually faster than transactions made with personal checks. Consider the length of time required for a clerk to write information on the back of a check, as opposed to swiping the card through a terminal and entering the purchase amount. However, if debit card transactions cause delays in checkout or other forms of inconvenience to the customer, the merchant will no doubt abandon the debit card.

From the bank's point of view, the debit card offers convenience to customers and provides a less expensive way to process deposit-related transactions. A bank's identity is also enhanced when cards are presented to merchants. However, most customers enjoy the delayed payment schedule credit cards have to offer. As a result, bankers will have to improve their sales techniques if they want to persuade customers to access personal deposits for their purchases.

The prospects for growth of debit cards in this country are excellent. Proprietary debit card transactions in particular are growing through regional networks. Some major oil companies accept debit cards at automated gasoline pumps. Although some pricing and operational issues are yet to be resolved, consumer and merchant acceptance are increasing.

Private Label Card

The private label card, as it applies to banking, is another outgrowth of the conventional bank credit card. The concept, however, has its roots in the retail industry.

Long before the bankcard, retailers introduced credit plans to allow customers to "buy now and pay later." Customers would apply for credit to make large purchases, such as washing machines or refrigerators. As credit plans evolved, retailers saw opportunities to stimulate sales by offering credit for smaller purchases as well. Additional profits could be made through charging interest on the money borrowed by customers for purchases, especially when the cost of borrowing money to fund the credit portfolio was relatively low and stable. Eventually, the retail "charge card" became a common vehicle for customers to make credit purchases. This card was uniquely tied to the retailer issuing the card and could be used only in that retailer's stores.

This one-to-one credit relationship between retailer and customer underlies the concept of the private label card in banking today: the card can be used only in stores designated by the name or label of the retailer that appears on its face. This, of course, distinguishes the private label card from the conventional bankcard, which can be used at millions of merchant locations worldwide.

The notion of a private label card in banking started to gain popularity during the 1970s. The bank private label card is similar to the retail charge card: the private label card can be used only in stores of the retailer named on the card. The key difference is that credit for the private label card is extended by a bank, which has a contractual agreement with the retailer. In most cases, the bank approves the application, produces the card, processes customer billings and payments, and handles customer service activities associated with the credit and processing functions.

A cousin of the private label card, the affinity group card, gained acceptance during the first half of the 1980s. Some bankers see it as another approach to marketing the bankcard. The affinity group card is a MasterCard or Visa card designed for groups with some form of common interest or relationship. Examples include cards for professionals (physicians, lawyers, teachers, and so on), alumni of a specific college or university, members of retired persons organizations, or military personnel. The graphic design of the card is tailored for the specific affinity group. A pilots association, for instance, might receive a card that graphically represents some recognizable aspect of flying, such as a pilot's wings, on the face of the card.

Private label benefits to the consumer must be created by the merchant and the bank. For example, the cardholder may be given special discounts or rebates on merchandise, receive special mailings for sales discounts (such as "moonlight" or one-day sales), or receive purchase credits. In some cases, the private label card may be the only credit card that can be used in the store. Affinity groups may receive special travel packages or merchandise. These broad-based programs might be offered to affinity groups because members may use their cards with any merchant that accepts MasterCard or Visa.

The private label card offers some specific advantages to the merchant. First, the possibility of repeat sales is enhanced because the card can be used only with the designated merchant. Second, sales promotions can be directed at specific consumers, who, as cardholders, should be more likely to respond. Third, the costs of funding and processing the portfolio are avoided. Finally, the merchant may be able to negotiate a more favorable credit approval arrangement with the bank so that more applicants are approved to receive cards. This improves the chances for additional sales through increased floor traffic (more shoppers entering the store) and a larger target group for promotional efforts.

During the 1970s, bankers became increasingly concerned that the consumer market for conventional bank credit cards was approaching the saturation point.

Private label cards presented a fresh avenue for growth of credit card portfolios and a new means to establish banking relationships with more consumers. A few banks set up special departments for private label merchants that would send statements with the retailer's name printed conspicuously on the front. Some banks even had employees in their private label departments answer telephone calls using the retailer's name.

While the private label card yields rewards for the bank, it also carries risks. Pricing the card to the cardholder and pricing the service to the merchant can be complex and require careful structuring. For example, if the merchant favors more consumer credit approvals, the bank may take a higher level of loan loss risk into the private label card portfolio. The consequences could be more charge-offs and lower profits for the bank.

Both the merchant and the bank should share the risks and rewards. This may take the form of a merchant reserve account (deposit) to absorb a higher level of loan loss. It might also consist of a direct reimbursement from the merchant to the bank if losses exceed an agreed level. Any private label agreement must mutually benefit merchant and banker.

There appear to be opportunities for growth of private label card programs in banks. But any bank considering this type of program must recognize the difference between the conventional bankcard portfolio and the private label portfolio. Private label portfolios generally have lower credit limits, fewer fees, and a higher percentage of minimum payments. The credit criteria are usually less restrictive, while delinquencies and charge-offs are higher relative to the total size of the portfolio. The bank wants higher outstanding balances for interest income, while the retailer wants more credit available for sales. Therefore, banks must carefully monitor their portfolio performance, administrative and processing expenses, and overall program profitability.

Smart Card

The smart card, also known as the chip card, uses a different technology from that used for conventional bankcards. Pioneered in France, the smart card contains a computer chip, embedded within the plastic, that provides unique capabilities. Using special terminals designed to interact with the embedded chip, the card can perform special functions. For example, if an unauthorized person tries to use the card at a POS terminal and inputs an incorrect customer identification code, the chip self-destructs and renders the card useless.

In addition, data can be stored on the chip so that transactions must be carried out within prescribed limits. For instance, the available credit can be reduced each time a purchase is made with the card. If the purchase amount exceeds the credit available (a running total is stored on the chip), the purchase will not be authorized. After the cardholder makes a payment each month, the amount of available credit is increased.

Although the chip card uses advanced technology and offers numerous benefits to the customer, the merchant, and the bank, several significant issues revolve around it. In the United States, ATMs and other POS terminals currently use magnetic stripe technology. Substantial investment would be required to convert thousands of magnetic stripe terminals to interact with chip cards. Unless an economically feasible means for integrating chip and magnetic stripe technologies is identified and agreed on, the chip card will be slow to gain acceptance in the United States.

Summary

When we think of bankcards, bank credit cards immediately come to mind. However, some other convenience cards are available on the market, such as ATM cards, check guarantee cards, debit cards, department store charge cards, and smart cards. These cards have different functions and are unlikely to rival the mass appeal of credit cards.

Review Questions

1. How is the debit card different from the conventional bank credit card?

2. For merchants, what are the main attractions of the private label card?

3. How do affinity group cards differ from private label cards?

4. What are some possible chip card applications outside conventional financial transactions?

A Final Word

This text has attempted to give you an understanding of the growth, development, and mechanics of the bank credit card business. As it has evolved from the 1960s to the early 1990s, the card has fundamentally changed consumers' shopping and purchasing habits. It has changed the way banks and other financial institutions deliver and control consumer lending. It has opened additional avenues of revenue to the merchant and shifted retail credit risk to card issuers. No one would have predicted the phenomenal growth that bankcards have enjoyed over the last 25 years.

Given the growth to date, what can we expect the future to hold for the bankcard business? To appreciate the pitfalls of predicting the future, we need only look at the events of recent history. Since work began on this book, the people of the USSR have rejected communism, and the government has outlawed the Communist Party. The Berlin Wall has been knocked down, uniting East and West Berlin. A year ago, most people would not have predicted such rapid, radical change. With that caveat in mind, it is still safe to predict that a number of trends will continue in the bankcard industry.

♦ Bankcards will continue to grow in importance as both a credit and payment device.

♦ The consolidation of issuers and acquirers will continue, both through bank mergers and acquisitions and the sale of bankcard portfolios.

♦ The polarization of issuers and acquirers will continue.

♦ More nonbank entities will enter the business, either in partnership with existing banks or by forming their own banks.

♦ There will continue to be pressure on bankcard profitability:

  ◊ AT&T's introductory "no-fee-for-life" offer best exemplifies the pressure on annual membership fees.
  ◊ The move to a variable interest rate structure by many banks will impair profitability.
  ◊ The high cost of enhancements designed to distinguish card issuers from one another can only compress profits over time.
  ◊ The entrance of nonbank entities, whose main goal is to protect their core business, will tend to lower profitability.

Globalization of the card business will present significant opportunities in Europe, Eastern Europe, and Japan.

Survival, Growth, and Profitability

The critical issues for the decade of the nineties are survival, growth, and profitability. The key to survival will be quickness of action—the ability to recognize an opportunity and act upon it. The key to growth is research—demographic, psychographic, and customer communication. Understanding the customer's needs and wants is critical. Obviously, the key to profitability is survival and growth. However, there must also be a commitment to service, value, and flexibility. There must be a determination to forgo short-term gain for long-term strategic benefit.

Glossary

access card The card used in an automated teller machine for deposits, cash withdrawals, account transfers, and other related functions.

account history The payment history of an account over a specified time, including the number of times the account was past due or over the credit limit.

acquirer In interchange, an institution that maintains the merchant relationship and receives all transactions.

acquirer's interchange discount (AID) In interchange, a bank that maintains the merchant relationship and receives all transactions from the merchant.

acquiring bank In interchange, a bank that maintains the merchant relationship and receives all transactions from the merchant.

affinity group A collection of individuals with some form of common interest or relationship, such as professions, alumni, retired persons organizations, etc.

agent bank A bank that, by agreement, participates in another bank's card program, usually by turning over its applicants for bankcards to the bank administering the program and by acting as a depository for merchants.

aging The procedure by which accounts are classified for the purpose of determining delinquency, ranging from a current to a charge-off status.

altered card A card on which the original account number has been changed to allow fraudulent use.

altered sales draft A sales draft in which the dollar amount was changed to read something other than what the cardholder actually agreed to and signed.

annual percentage rate (APR) The amount of finance charge applicable to outstanding balances for one year. This is frequently expressed as the periodic (monthly) rate x 12 = the APR.

approval ratio The number of cardholder applications approved, expressed as a percentage of total applications received.

asset and liability policies Policies that govern the bankwide amounts and funding sources for various loan classes in the bank's total loan portfolio, as well as the mix of liability classes for loan funding and investment purposes.

ATM *See* automated teller machine.

---

attrition The loss of accounts either involuntarily, because of bad debts, death, etc., or voluntarily, at the option of the cardholder.

authorization Approval by, or on behalf of, the card issuer to validate a transaction for a merchant or another affiliate bank.

authorization request A request for approval by the merchant to validate a cardholder sales transaction.

automated teller machine An unmanned electronic device that performs basic teller functions, such as accepting deposits, cash withdrawals, account transfers, loan payments, and account balance inquiries.

available credit The difference between the credit limit assigned to a cardholder account and the present balance, including authorizations outstanding (for which the actual transaction has not yet been received through interchange).

average balance The average amount outstanding on a cardholder account within a specified period of time. This is calculated by adding all the monthly balances on an account within a specified time period and dividing the total by the number of months within that time period.

average daily balance An amount calculated by dividing the balance outstanding at the close of each day during the billing cycle by the number of days in that cycle.

back-end processing Refers to activities that do not involve customer contact or risk management, for example, authorization and cardholder billing.

behavioral score A score that predicts the account behavior most likely to occur according to statuses ranging from a current status to a charge-off status.

billing date The month, day, and year when a periodic (or monthly) statement is generated. The billing date is made when calculations for appropriate finance charges, minimum payment due, and new balance have been performed.

card issuer The financial institution that authorizes issuing a bankcard for which the institution or its agent carries the liability.

card mailer A carrier used in mailing a card to the cardholder. The mailer may contain instructions regarding the conditions of card use.

card reissue The process of preparing and distributing bankcards to cardholders whose cards have expired or will soon expire.

cardholder agreement The written understanding stating the terms and conditions of using a card.

cardholder bank The bank that has issued a bankcard to an individual. The term is frequently used in interchange arrangements to identify the card-issuing bank.

cardholder masterfile The bank record of all cardholder accounts, including all information pertinent to the accounts (name, address, credit limit, payment history, etc.).

cash advance A cash loan obtained by a cardholder through presentation of a card at a bank office, at an ATM, or by mail request.

central processing unit (CPU) The main computer processor.

characteristic A specific item of information (such as income) on a credit application used in the credit-scoring method.

chargeback A transaction that is challenged by a cardholder or merchant bank and sent back through interchange to the bank of account (cardholder or merchant) for resolution.

charge-off (1) The balance on a bank cardholder account that a bank no longer expects to be repaid and writes off as a bad debt. (2) The process of charging off accounts. A charge-off is generally recorded by a debit to the reserve for possible credit losses and a credit to the loan balance. *Also called* bad debt.

check guarantee A service provided through a plastic card that guarantees payment up to a defined limit, provided that the merchant follows proper steps in accepting the check.

chip card *See* smart card.

collateral Specific property, securities, or other assets pledged by a borrower as a backup source of loan repayment.

contingent liability The total amount of credit available to borrowers, but not in use, at any period in time.

convenience user A cardholder who pays the balance in full on each due date.

corporate card A bankcard issued to companies for use by company employees. The liability for abuse of the card typically rests with the company and not with the employee.

country club billing A billing system in which copies of transactions are mailed to the cardholder with a monthly statement.

credit analyst A person who performs the function of reviewing applications for credit, together with pertinent credit data, and decides to approve or decline individual applications.

credit balance The amount of credit available to a cardholder after the credits and debits have been posted in a billing cycle.

credit criteria The standards applied to cardholder applications or to previous account records in order to determine approval or declination of the application for a credit card, a line of credit, or an increase in the line of credit.

credit limit The amount of the credit line set by the card issuer for the cardholder's account.

credit loss The amount lost (charged off) as a result of the failure of the cardholder to repay the amount owed on the account.

credit scoring A method for predicting the creditworthiness of applicants for credit.

cross-sell The use of one product or service as a base for selling additional products and services.

current account A cardholder account that has been paid up to date and on which no amount is past due.

cutoff score Under the credit-scoring method, the minimum score an application must carry to be approved.

cycle The grouping of cardholder accounts to provide for a distribution of workload and easier account identification.

cycle period A specified period during which both debit and credit transactions are accumulated for billing.

data capture A data processing term for the collection, formatting, and storage of data in computer memory according to predefined fields, for example, customer name, account number, and dollar amount of purchase. When a terminal reads this information from a plastic card or from entries at a terminal, the information is stored in computer memory for later output as a hard copy printout or as soft copy on a CRT display.

data entry *See* data capture.

DBA (doing business as) Refers to the specific name and location of the merchant's store where a bankcard purchase is made.

debit A charge to a customer's bankcard account.

debit card A plastic card, issued by a financial institution that charges the customer's personal account. The card may be proprietary (issued solely by one institution), or it can be a regionally or nationally accepted card.

delinquent account An account on which payment has not been made according to the terms and conditions of the cardholder agreement; an account on which payments are past due.

derogatory information Data received by a lender indicating that an applicant or cardholder has not paid his or her accounts with other creditors.

descriptive billing A method of billing in which each monetary transaction posted to an account during a billing period is identified and described on a bill.

dial-up terminal A merchant authorization device that, like a telephone, dials the authorization center for validation of transaction.

direct expense Any expense directly attributable to operations, including operating expense, advertising and other marketing expense, cost of funds, loan and fraud losses, and the like.

discharge Under the Bankruptcy Code, an action that releases the debtor from any legal obligations to repay debts.

disclosure Information required by law to be given to cardholders relative to the terms of the credit extended. Disclosures must appear on cardholder agreements, monthly (periodic) billing statements, or any documents in which finance charge rates are mentioned.

discount rate The fee a merchant bank charges the merchant for giving the merchant deposit credit and handling the merchant's sales drafts or electronic sales transactions.

dual dating The practice of embossing two dates on the face of credit cards with the first date as the effective date and the second as the expiration date. The purpose is to deter the fraudulent use of cards.

duality The handling of both MasterCard International and Visa International transactions (cardholder, merchant, or both) by a bank.

effects test A test of credit criteria to see if they have the effect of discriminating on the basis of any of the prohibited classifications under the Equal Credit Opportunity Act.

electronic funds transfer system (EFTS) An electronically based system designed to eliminate the paper instruments that are normally associated with the movement of funds, such as a cash withdrawal from an ATM that eliminates the necessity of writing and processing a check.

emboss The process of printing identifying data on a bankcard in the form of raised characters.

encoding The magnetized recording of data on the magnetic stripe on the bankcard.

error resolution The process of resolving mistakes made after incorrectly billing customers for transactions. Regulation E outlines the requirements for this process.

exceptions Transactions that fail to meet the parameters of the system.

float Money balances that appear for a period of time on both the balance statements of payer and payee due to a lag in the collection process.

floor limit The amount of purchase over which the merchant must obtain a transaction authorization by telephone or authorization terminal.

funds matching A financial term used to describe the funding strategy that uses short-term (generally interest-sensitive) deposits to fund short-term (generally interest-sensitive) loans, long-term funds for long-term loans, etc.

grace period The period of time between the statement date and the payment due date in which no interest is charged if the balance due is paid in full.

gross income The total dollar amount of all income sources for a credit card operation.

hot card A card used on an account on which excessive purchasing is taking place. This may indicate a lost or stolen card or other unauthorized purchasing.

imprinter A device supplied to the merchant to produce an image of the embossed characters of the bankcard on all copies of sales drafts and credit slips.

incoming chargeback *See* merchant chargeback.

incoming interchange Transactions from cardholder activity received from the MasterCard International and Visa International networks by the cardholder bank.

interchange The domestic and international systems operated by MasterCard International and Visa International for authorization, settlement, and the passing through of interchange and other fees, as well as other monetary and nonmonetary information related to bankcard activities.

interchange fee The amount paid by the merchant bank to the cardholder bank on each sales transaction. MasterCard International and Visa International independently establish interchange fees for their networks.

issuer A member of an interchange system that issues cards (for example, a bank that belongs to MasterCard International and issues MasterCard cards). Within a transaction interchange network, the institution issues and verifies identification and authentication information on a customer.

late charge A financial penalty against the cardholder for failure to make the minimum payment. Normally imposed 15 days after the payment due date.

laundered drafts The mingling of fraudulent drafts with legitimate drafts in an attempt to hide the fraudulent activity.

line of credit The amount of credit a lender will extend to a borrower over a specified period of time.

loss reserve The amount on the balance sheet set aside to pay for projected credit losses from the cardholder portfolio.

magnetic ink character recognition (MICR) The machine recognition of characters printed with ink containing particles of a magnetic material. In bankcard operations, this is used most frequently with merchant draft processing, MICR drafts and remittance processing, and MICR checks and remittance coupons.

magnetic stripe A stripe of magnetic information affixed to the back of a plastic credit or debit card. The magnetic stripe contains essential customer and account information.

masterfile A computer file composed of records with similar characteristics or containing data of a relatively permanent nature. A cardholder masterfile would contain such information as names, account numbers, addresses, credit limits, expiration dates, and number of cards issued as minimum data.

merchant accounting The recording by a bank of the number and dollar value of all sales drafts and credit slips submitted by each merchant.

merchant agreement A written agreement between a merchant and a bank containing their respective rights, duties, and warranties with respect to acceptance of the bankcard and matters related to bankcard activity.

merchant assessment *See* discount rate.

merchant authorization The means of receiving sales validation for the merchant, by telephone or authorization terminal, to guarantee payment to the merchant.

merchant bank The bank that has entered into an agreement with a merchant to accept deposits generated by bankcard transactions.

merchant chargeback A transaction that is challenged by a cardholder bank against a merchant bank and comes to the merchant bank through interchange. Also referred to as incoming chargeback.

merchant file A computer record of information on all merchants serviced by a merchant bank.

merchant number A series or group of numbers that numerically identifies each merchant to the merchant bank for accounting and billing purposes.

monetary transaction Any transaction posted to an account that has a dollar value.

national association MasterCard International or Visa International, which are licensing and regulatory agencies for bankcard activities.

national debit card A card that requires compliance with all the operating regulations of MasterCard International and/or Visa International to be used in interchange.

negative file A record containing all accounts for which charge privileges have been revoked by the card issuer.

net charge-off The gross dollar amount charged to bad debt (or loan loss), less recoveries received, during a specified period.

net settlement The net amount settled after the values of cardholder purchases and merchant sales have been determined in interchange.

nominal percentage rate The disclosed interest rate, expressed as a rate for a 12-month period.

nonmonetary transaction Any transaction posted to an account that does not have a dollar value affecting the account. Name, address, and status changes are examples of nonmonetary transactions.

nonsufficient funds (NSF) The designation given to checks that are returned because there was not enough money in the cardholder's account to cover the check when it was presented to the bank for payment.

off-line An operating mode in which terminals (or ATMs) are not connected to a central computer source for an extended or brief period. Responses are governed by the parameters, or guidelines, set within the terminal or supporting device as defined by the card issuer. Information is not accessible in a live environment, meaning that current active files are not able to be viewed as a transaction is conducted.

online An operating mode in which terminals (or ATMs) are connected to a central computer system and have access to the database for authorization, inquiry, and file changes. Live files are accessed for each transaction.

open-to-buy *See* available credit.

optical character recognition (OCR) Electronic reading and digital conversion of numeric or alphabetic characters from printed documents. In bankcard operations, this is used most frequently for merchant draft processing, card production, and remittance processing.

optical scanning *See* optical character recognition.

outgoing chargeback A transaction that is challenged by a cardholder bank against a merchant bank and sent out to the merchant bank through interchange.

over-limit account An account on which the credit limit has been exceeded.

P&L The profit and loss statement of the bank. Also referred to as the operating statement.

payment coupon The section of the billing statement containing payment information, which should be returned by the customer with the payment.

payment due date The date by which payment must reach the bank to keep the account in a current status.

periodic rate An amount of finance charge, expressed as a percentage, which is to be applied to an appropriate balance for a specified period, usually monthly, providing there is a balance that is subject to a finance charge.

PIN (personal identification number) The individual number or code uniquely assigned to customers for identification when their cards are used in ATMs or terminals at the point of sale.

point-of-sale (POS) system An electronic system that accepts financial data at or near a retail selling location and transmits that data to a computer or authorization network for reporting activity, authorization, and transaction logging.

POS terminal A device placed in a merchant location that is connected to the bank's system via telephone lines and is designed to authorize, record, and forward data by electronic means for each sale.

positive authorization An authorization procedure in which every account on file in the computer system can be accessed to determine its status before an authorization is granted or declined.

positive file A file containing, at a minimum, the current balance for each active cardholder account. A positive file may also include PIN and other cardholder information.

posting The process of recording debits and credits to individual cardholder account balances.

presentment A demand made by the holder of a sales draft to the merchant bank asking for acceptance or payment of a credit card purchase.

prime rate The rate a bank charges its best commercial customer to borrow funds.

primary account number (PAN) The embossed and/or encoded number that identifies the card issuer to which a transaction is to be routed and the account to which it is to be applied unless specific instructions indicate otherwise. The PAN consists of a major industry identifier, issuer identifier, individual account identifier, and check digit.

private label card A bankcard that can be used only in a specific merchant's stores.

proprietary debit card A bankcard designed for the exclusive benefit of customers of the issuing institution.

re-aging Bringing an account from a delinquent status to a current status, or keeping a delinquent account in its present stage of delinquency so that it does not advance in the cycle.

recourse The right to collect from a maker, seller, assignor, or endorser of an instrument or installment credit obligation if the first party liable fails to meet the obligation.

recovery function The activity involved in attempting to collect the balance owed after an account has been charged off.

reference number The number assigned to each monetary transaction in a cardholder billing system. Each reference number is printed on the monthly statement to aid in the retrieval of the document should the cardholder question it.

regional network A network that processes transactions for financial institutions and retailers in a given geographic area. They are not part of the national interchange system.

reissue The process whereby cardholder accounts, on which bankcards have expired or will expire, are issued new cards.

re-presentment A subsequent demand for acceptance of payment made after a chargeback has been disputed.

response time The length of time required to complete an electronic transaction, such as a file inquiry or file input.

restricted account A cardholder account to which a status code has been posted indicating that the cardholder may not use the account.

restricted card list A listing of cardholder accounts, in either alphabetic or numeric sequence, on which transactions are restricted and not to be completed by merchants without authorization.

revolving line of credit An account the customer can repeatedly use and pay back without having to reapply every time credit is used.

rollover The carrying forward of a portion of an outstanding balance on a cardholder account from month to month.

runaway card A cardholder account that has exceeded the limit for excessive purchases, may also have exceeded its credit limit, or continues to indicate other unauthorized use.

scrip A paper money substitute for purchases that is redeemable at participating merchant outlets at face value.

settlement The process by which merchant and cardholder banks exchange financial data and value resulting from sales transactions, cash advances, merchandise credits, and the like.

skip account An account of a cardholder with a balance due whose whereabouts are unknown.

skip tracer An individual who is part of the collection function and attempts to find delinquent cardholders who have moved in order to avoid collection attempts and repayment of the balance owed.

smart card A card that carries an embedded computer chip with memory and interactive capabilities so that it can be updated.

spread The difference between a bank's cost of funds and its interest yield; a source of bankcard profitability.

standards integration A process of editing performance and quality standards to ensure that they complement, as opposed to conflict with, each other.

"statused" account A cardholder account to which a status code has been assigned indicating a condition under which the cardholder may not use the account.

suspense account A special classification for holding transactions until a problem is resolved. Suspense accounts must be funded by the bank on the balance sheet and therefore must bear a funding cost.

switch An electronic mechanism that routes transaction data from a POS terminal to the authorizing data processor for approval of the card-issuing institution. The switch can also serve as an authorization center for financial institutions whose accounts are maintained on another computer in the network.

T&E card "Travel and Entertainment" card that typically requires payment in full each month. Examples are American Express, Diners Club, and Carte Blanche.

unsecured credit Credit extended without recourse to attach specific assets of the borrower in the event of default.

usury (1) A higher rate of interest than is allowed by law. (2) The act of charging a higher rate of interest for the use of funds than is legally allowed by a state.

warning bulletin *See* restricted card list.

white plastic fraud Instances in which dishonest merchants create false drafts on real cardholder account numbers to deposit in merchant banks and withdraw cash.

yield The annual rate of return on an investment. Yield, in regard to bank cards, refers to the cardholder portfolio.

zero floor limit Requires that all cardholder transactions be authorized.

Index

The Bankcard Business   © 1992

Thank you for using this American Bankers Association/American Institute of Banking textbook. Your responses on the following evaluation will help shape the structure and content of future editions. <u>Return your completed form to your instructor or fold in three and mail to:</u> American Institute of Banking, Attn: Manager, Product Development, 1120 Connecticut Avenue, N.W., Washington, D.C. 20036.

Name of Chapter _____

Name of Bank _____

TEXTBOOK/COURSE ATTRIBUTES

Importance Factor					Satisfaction Level			
Very Important			Not Important		Completely Satisfied			Not Satisfied
1	2	3	4	Textbook covered all important topics	1	2	3	4
1	2	3	4	Content was easy to read and understand	1	2	3	4
1	2	3	4	The graphics and examples were helpful	1	2	3	4
1	2	3	4	I can use what I've learned in this course in my work	1	2	3	4

	Excellent			Poor
What was your overall opinion of the textbook?	1	2	3	4

Did your instructor use any additional materials to teach this course?

( ) Yes   ( ) No

If Yes, please check all that apply

( ) Transparencies/Overheads   ( ) Handouts

( ) Other textbook (please specify)_____

Number of AIB courses you have taken in past three years:

( ) 0   ( ) 1-2   ( ) 3-5

( ) More than 5

AIB course taken through:(Please check all that apply)

( ) AIB Chapter/Study Group

( ) AIB Correspondence Study Program

( ) Other (please specify)_____

Currently working toward an AIB Diploma/Certificate?

( ) Yes   ( ) No

If yes, please specify:

( ) Bank Operations   ( ) Consumer Credit

( ) Commercial Lending   ( ) General Banking

( ) Mortgage Lending   ( ) Accelerated Banking

( ) Customer Service   ( ) Securities Services

( ) Supervisory

Asset size of your bank:

( ) 0-$75m   ( ) $76-$250m   ( ) $251-$500m

( ) $501-$1b   ( ) over $1b

Number of employees in your bank:

( ) 1-10   ( ) 11-20   ( ) 21-40

( ) 41-90   ( ) 91-200   ( ) 201-350

( ) 351-2,000   ( ) over 2,000

Job Title:_____

Major Job Responsibility:

( ) Lending   ( ) Marketing   ( ) Operations

( ) Compliance   ( ) Auditing   ( ) Human Resources

( ) Trust   ( ) Customer Svc.   ( ) Branch Admin.

( ) Securities Processing   ( ) Security/Risk Management

( ) Other (Please specify)_____

Years in Banking:

( ) 0-2   ( ) 3-5   ( ) 6-10

( ) Over 10

Highest Education Level:

( ) High School   ( ) Some College   ( ) BA/BS Degree

( ) Advanced Degree

Age:

( ) under 25   ( ) 25-35   ( ) 36-45

( ) over 45

Name_____

Bank_____

Address _____

City _____ State _____ Zip _____

Telephone (____) _____

( ) Please send me more information on AIB's Diploma/Certificate Program.

( ) Please send me more information on AIB's Correspondence Study Program.

Comments (please identify any specific suggestions you have that may improve the overall effectiveness of this publication):

_____

_____

_____

_____

_____

_____

_____

_____

_____

_____